TRANSPORT MECHANISMS IN MEMBRANE SEPARATION PROCESSES

The Plenum Chemical Engineering Series

Series Editor
Dan Luss, *University of Houston, Houston, Texas*

COAL COMBUSTION AND GASIFICATION
L. Douglas Smoot and Philip J. Smith

ENGINEERING FLOW AND HEAT EXCHANGE
Octave Levenspiel

REACTION ENGINEERING OF STEP GROWTH POLYMERIZATION
Santosh K. Gupta and Anil Kumar

TRANSPORT MECHANISMS IN MEMBRANE SEPARATION PROCESSES
J. G. A. Bitter

A Continuation Order Plan is available for this series. A continuation order will bring delivery of each new volume immediately upon publication. Volumes are billed only upon actual shipment. For further information please contact the publisher.

TRANSPORT MECHANISMS IN MEMBRANE SEPARATION PROCESSES

J. G. A. Bitter

University of Twente
Enschede, The Netherlands

SPRINGER SCIENCE+BUSINESS MEDIA, LLC

Library of Congress Cataloging-in-Publication Data

Bitter, J. G. A.
Transport mechanisms in membrane separation processes / J.G.A. Bitter.
p. cm. -- (The Plenum chemical engineering series)
Includes bibliographical references (p.) and index.
ISBN 978-0-306-43849-3 ISBN 978-1-4615-3682-6 (eBook)
DOI 10.1007/978-1-4615-3682-6
1. Membrane separation. I. Title. II. Series.
TP248.25.M46B58 1991
660'.2842--dc20 91-4005
CIP

ISBN 978-0-306-43849-3

Originally published by Plenum Press, New York in 1991

To my dear wife Adrienne
for her encouragement and
support.

— Joop

PREFACE

The present book contains a comparison of existing theoretical models developed in order to describe membrane separation processes. In general, the permeation equations resulting from these models give inaccurate predictions of the mutual effects of the permeants involved, due to the simplifications adopted in their derivation.

It is concluded that an optimum description of transport phenomena in tight (diffusion-type) membranes is achieved with the "solution-diffusion" model. According to this model each component of a fluid mixture to be separated dissolves in the membrane and passes through by diffusion in response to its gradient in the chemical potential.

A modified Flory–Huggins equation has been derived to calculate the solubility of the permeants in the membrane material. Contrary to the original Flory–Huggins equation, the modified equation accounts for the large effect on solubility of crystallinity and elastic strain of the polymer chains by swelling.

The equilibrium sorption of liquids computed with this equation was found to be in good agreement with experimental results. Also, the sorption of gases in both rubbery and glassy polymers could be described quantitatively with the modified Flory–Huggins equation without any need of the arbitrary Langmuir term, as required in the conventional "dual-mode" sorption model. Furthermore, fewer parameters are required than with the conventional equations in order to achieve at least identical accuracy.

A new equation has been derived for predicting diffusion rate, according to which the self-diffusivity of the components in the polymer–permeants mixture must be applied instead of the mutual diffusivities used in the conventional equations. The self-diffusivity of the compounds in multicomponent mixtures is described accurately by a modified kind of Vigne equation, corrected for nonideal mixture behavior and extended for multicomponent mixtures. Procedures are given for estimating the basic binary diffusivities and basic self-diffusivities required to calculate mixture diffusivities in polymers.

The "solution-diffusion" model has been employed in order to derive a new permeation equation, which accounts for nonlinear concentration gradients inside membranes. This equation is capable of predicting the mutual effect of the permeants on their permeation rates. Fluxes and selectivities computed with this equation from thermodynamic and physical parameters are in good agreement with experiments.

J. G. A. Bitter

Enschede, The Netherlands

ACKNOWLEDGMENTS

I wish to express my gratitude to the Koninklijke/Shell-Laboratorium, Amsterdam (Shell Research B.V.) for permitting publication of the contents of this book. The main part of this work was researched and developed in this laboratory.

I also thank my colleagues from the Separation Technology Department of K.S.L.A., who were involved in these investigations, for their valuable contributions.

Many of the drawings were prepared by my good friend Jim Edwards from Princeton (N.J.).

I am grateful to Prof. Dr. C. A. Smolders of the University of Twente and Prof. Ir. J. H. Wesselingh of the University of Groningen in The Netherlands for the extensive discussions I had with them. They led to a better understanding of the subjects treated in this book.

CONTENTS

CHAPTER 5. BASIC DIFFUSION EQUATION

CHAPTER 6. SOLUBILITY OF PERMEANTS IN SEMI-CRYSTALLINE AND CROSSLINKED POLYMERS

CHAPTER 7. COMPARISON AND EXPERIMENTAL CHECK OF THE SOLUBILITY EQUATIONS

CHAPTER 8. PREDICTION OF DIFFUSIVITY IN MULTI-COMPONENT MIXTURES

CHAPTER 9. NEW PERMEABILITY EQUATIONS

APPENDIX I. IRREVERSIBLE PROCESSES NEAR EQUILIBRIUM

APPENDIX II. DERIVATION OF THE CONCENTRATION POLARIZATION EQUATION OF GASES

APPENDIX V. THEORY OF THE ENTROPY OF MIXING

1

INTRODUCTION

Membrane separations represent a new type of unit operation which, ultimately, is expected to replace a significant proportion of conventional separation processes. The advantage of membrane separations lies in their relatively low energy requirements, the reason for which is that, unlike conventional processes such as distillation, extraction, and crystallization, they generally do not feature phase transitions. If, however, in a membrane process a phase transition does occur (as in pervaporation), this advantage does not exist, and economically attractive applications are restricted to special cases. In gas separations, which in principle do not involve phase transitions, the energy consumption in a membrane process is in general no less than that, for instance, in an absorption–desorption process. In fact the two processes are similar, the membrane corresponding to the absorbing medium. This is insufficiently recognized by scientists and process engineers working with membranes.

Thus membrane separation is not an ideal new technology, but just another unit operation, whose attractiveness must be weighed against that of other competitive processes.

For evaluating the advantages and disadvantages of membrane processes it is essential to have at one's disposal a reliable physical model that permits a sufficiently accurate estimation of the technical and economic feasibility. Such a model also enables optimization of the separation processes and the development of new and suitable membranes and membrane modules.

It is the aim of this book to explain and compare models previously developed for describing the transport mechanisms involved in membrane separation systems. In doing this, we place emphasis mainly on the "solution-diffusion" model, which in our opinion represents a correct description of separation processes with tight membranes. The various existing diffusivity equations are discussed and we derive a modified diffusivity equation, which serves as the basis for the derivation of a new general membrane permeation equation. Procedures are given for determining and estimating the various parameters required to calculate the fluxes and selectivities from the new membrane permeation equation.

The main parameters of the "solution-diffusion" model are the solubility and diffusivity of the permeants in the membrane. For the solubility of liquids in a polymer the Flory–Huggins equation is usually applied. Owing to the simplifications made in its derivation, the accuracy of predictions is generally not good enough.

For describing solubility of gases in polymers a distinction is made between "rubbery" and "glassy" polymers. Gas solubility in the former type of polymer is generally expressed by the linear Henri's law. For describing gas solubility in "glassy" polymers a more complicated nonlinear equation is needed, for which usually the arbitrary "dual-mode" sorption equation is used. With this equation an accurate description of gas sorption is achieved, but it contains a large number of empirical parameters that limit its practical application.

As mentioned before, a distinction is made between gases and liquids when describing solubility in polymers. However, after sorption into the polymer the permeants are in such close contact with each other and with the polymer chains that they must be regarded as liquids, irrespective of their original phase condition. Therefore, it should be possible in principle to express the solubility of gases and liquids in polymers by the same equation if the conditions inside the polymer are considered instead of those outside, as is common use in the conventional equations.

Such an equation is derived in this book and, furthermore, it accounts for the various shortcomings of the Flory–Huggins equation as well as those of the dual-mode sorption equation. We also derive a new equation for the diffusivity of a permeant in a multicomponent permeant–polymer mixture. In principle, it is based on the well-known Vigne equation for the diffusivity in binary mixtures.

2

TYPES OF MEMBRANE SEPARATION PROCESSES, MECHANISMS OF SEPARATION

Membrane processes may be classified according to the types of membranes used. With porous membranes, for example, a distinction is made between microfiltration and ultrafiltration, depending on the pore sizes and particle sizes involved. As an extension to this, reverse osmosis is sometimes called hyperfiltration and indeed theories have been developed, such as the "preferential sorption–capillary flow" model of Sourirajan,[1] in which separation is considered to take place via pores with dimensions in the range of molecular sizes.

Another common way of classifying the membrane processes is according to the exerted driving force. In nonporous (tight) membranes, which are used in, for example, reverse osmosis, dialysis, pervaporation, vapor permeation, and gas separation, different driving forces or phase conditions are applied (see Table 1).

Various models[2] suggest fundamental differences between the separation mechanisms of these processes in tight membranes. Such differences, however, exist only if we consider the conditions outside the membranes. Within the membranes the transport mechanisms are identical for the processes mentioned. This has been shown in theoretical studies of Cheng H. Lee[3] and the author[4] and has been confirmed experimentally by the latter with mixtures of hydrocarbons. Because of its importance this subject will be considered more closely in this book.

Table 1. Separation Processes with Tight (Polymeric) Membranes

Process	Driving force
Pervaporation	Gradient of vapor pressure
Vapor permeation	Gradient of vapor pressure
Gas permeation	Pressure gradient
Reverse osmosis	Pressure gradient
Thermoosmosis	Temperature gradient
Dialysis (osmosis)	Concentration gradient
Pertraction	Concentration gradient
Electrodialysis	Gradient in electric potential

2.1. POROUS MEMBRANES

In micropore membranes (pore sizes between about 200–3000 nm) viscous flow (Poiseuille flow) takes place under the influence of a pressure gradient as a driving force. The permeating molecules collide with each other, thus exerting mutually a frictional force. As a consequence all molecules pass the pores with the average drift velocity independent of their size, shape, or mass. Therefore, transport through microporous membranes is *nonseparative* on a molecular base for liquids as well as for gases.[5]

At pore sizes below about 5 nm, large molecules such as polymers can be retained from a solution. Then the process is called ultrafiltration. Apart from the "sieving" action of such membranes, separation of molecules by size and shape also takes place by steric hindrance at the entrance of the pores and by frictional resistance in the pores, both effects increasing with the ratio of the molecular diameter (a) to the pore diameter (d). Pappenheimer[6] took these effects into account and derived for the effective cross-sectional area of a pore (A_{ef}) the following expression:

$$A_{ef} = A_0\{(1 - a/d)^2\ [1 - 2.104(a/d) + 2.09(a/d)^3 - 0.95(a/d)^5]\} \quad (2.1)$$

in which A_0 is the real cross-sectional area of the pore. For mixtures of spherical molecules with molecular weights of 100 and 600, respectively, we calculated the relationship between permeate and retentate composition, depending on the cutoff molecular weight of the ultrafiltration membranes, assuming the permeation rate of the permeants is proportional to A_{ef} and to their volume fraction in the retentate. From the results, shown in Fig. 1, it is clear that, despite the large differences between the sizes of the molecules and the cutoff molecular weights of the membranes under consideration, selectivities of even 3 can be expected solely for geometrical reasons. By accounting also for the concentration gradients, caused by this selective transport, even larger selectivities are found.[7]

Apart from the previously mentioned separation mechanisms of porous membranes, use can also be made of "Knudsen flow" selectivity, if the pore diameters are smaller than the mean free paths of the molecules to be separated; or, if the surface-to-volume ratio of the pores is large enough to favor surface flow, preferential adsorption can be applied.

According to Budtov *et al.*,[8] viscous flow of gases is observed in capillaries at pore sizes above about 10^{-3} cm and "Knudsen flow" at pore sizes below about 10^{-6} cm. At intermediate pore sizes both types of flow occur simultaneously.

In the case of "Knudsen flow," the chance that a permeating molecule will collide inside the pore with another molecule is negligibly small with respect to the chance of it colliding with the pore wall. Hence the permeating molecule will pass that pore at its own molecular speed, which is inversely proportional to the square root of its molar mass.[5] Although the resulting "Knudsen" selectivity is generally low, this principle does have at least one interesting application: that of separating hydrogen from mixtures with hydrocarbons in, for example, the dehydrogenation process, in which porous ceramic membranes[9] are used. In this way the chemical

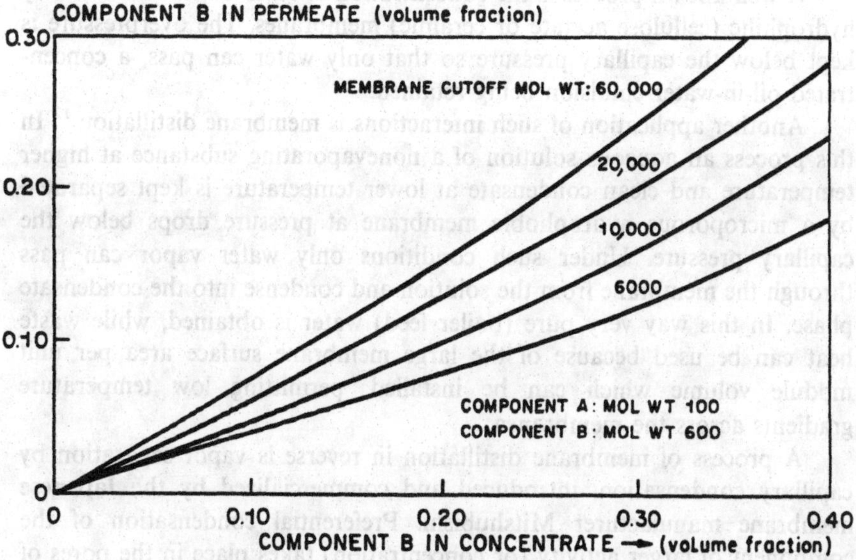

Figure 1. Molecular separation with porous membranes.

equilibrium is shifted toward higher conversions, thus reducing considerably capital and operating costs.

At pore sizes below, say, 3 nm, the surface area of the pores is so large with respect to the volume of the pores that the average residence time of a molecule on the pore wall becomes noticeable. Then at least part of the transport takes place along the pore surface. In such a process, which is called surface flow, transport of those compounds which are adsorbed preferentially is enhanced. For instance, in a reverse osmosis experiment performed on a hydrocarbon mixture, consisting of residual mineral oil, methyl ethyl ketone, and toluene (weight ratio 1:1:1) at 20°C, the selectivity in favor of the solvent mixture compared to the oil amounted to 20 with a Vycor glass membrane (pore size 3 nm),[10] while according to Eq. (2.1) a selectivity of about 3 would be expected. In this experiment preferential sorption of the solvents took place onto the strongly polar pore surfaces. On increasing the temperature to 80°C the selectivity dropped to below 3 on account of the decrease in preferential adsorption.

Specific interactions can also be utilized for breaking emulsions. A very strong emulsion of a concentrated aqueous H_2SO_4 solution (droplet size 10 μm) in an alkylate fraction, stabilized with a dispersant, can easily be broken by a microporous glass fiber membrane with an acrylic binder.[11] The acid phase coagulates onto the polar glass fibers and the oil phase onto the hydrophobic polymeric binder. Both phases were found to pass the membrane at an average flux of $10\ m^3/(m^2 \cdot bar \cdot day)$.

A well-known process is the concentration of oil-in-water emulsions by hydrophilic (cellulose acetate or ceramic) membranes. The overpressure is kept below the capillary pressure so that only water can pass, a concentrated oil-in-water emulsion being retained.

Another application of such interactions is membrane distillation.[12] In this process an aqueous solution of a nonevaporating substance at higher temperature and clean condensate at lower temperature is kept separated by a microporous hydrophobic membrane at pressure drops below the capillary pressure. Under such conditions only water vapor can pass through the membrane from the solution and condense into the condensate phase. In this way very pure (boiler feed) water is obtained, while waste heat can be used because of the large membrane surface area per unit module volume which can be installed, permitting low temperature gradients across the membranes.

A process of membrane distillation in reverse is vapor separation by capillary condensation, introduced and commercialized by the Japanese membrane manufacturer Mitshubishi. Preferential condensation of the component of larger activity (or concentration) takes place in the pores of the membrane. The condensed liquid in the pores is evaporated on the permeate side by evacuation. Mitshubishi claims that a selectivity of more than 200 in favor of water relative to ethanol can be achieved in this way.

The development of porous membranes, which was originally focused on polymeric materials, nowadays shows increasing application of ceramics. Total sales of ceramic membranes are expected to grow by 30% annually, for the most part at the expense of polymeric membranes.[13] The strong hydrophilic character of ceramics renders them eminently suited for treating aqueous solutions and emulsions. Their high thermal and chemical stability warrants their application at high temperatures and under severe corrosive conditions such as, for example, in dehydrogenation reactors for equilibrium shift.[9]

The possibility of handling mixed oxides is encouraging the development of catalytic ceramic membranes.[14] Catalytic activity can also be introduced by impregnation of ceramic membranes.[15] This technique can also be used to build-in active groups into the pore surfaces for enhancing preferential adsorption, thus improving the selectivity of membrane separation.

A quite new development is the preparation of membranes from molecular sieves, which potentially have extremely high selectivities. Preliminary investigations with "carbon molecular sieve membranes" have been performed by Koresh and Sofer.[16] They prepared such membranes by pyrolysis and thermal treatment of polymeric hollow fibers. High gas selectivities were indeed observed with some of these membranes; the reproducibility of the membrane production is still very poor, however.

The preparation of Zeolite membranes, although mentioned in the patent literature,[17] is still far from technical realization.

2.2. LIQUID MEMBRANES

Unsupported liquid membranes have been invented by N. N. Li.[18] In fact they consist of a very stable emulsion of an aqueous solution of a reagent and an immiscible hydrocarbon phase, the latter liquid being the continuous phase. By gentle stirring this emulsion is contacted with an aqueous phase, containing components which have to be removed. These components dissolve in the resulting oil drops and diffuse through to the contained aqueous solution droplets, where they are neutralized or converted by the reagent present. Sometimes a complexing agent is added to the oil phase. The process is then called mediated or facilitated transport; otherwise it is called nonmediated transport.

After reaction the reagent containing oily emulsion, which is stabilized by the addition of dispersants or emulsifiers, is separated from the aqueous phase and broken, for example by means of the aforementioned microporous membrane with hydrophilic and hydrophobic arrays,[11] in order to recover the various ingredients supplied.

In later developments microporous polymer membranes were impregnated with a liquid containing a complexing agent[19] that was immiscible with at least one of the contacting phases. This "supported liquid membrane" seems technically to be more feasible. The main problems are losses of the impregnation liquid into the contacting phase(s) (membrane instability), low permeant fluxes, sensitivity to overpressure due to irregular pore size distributions of the polymeric membranes, and fouling of the liquid phases with surface-active contaminants.

2.3. TIGHT MEMBRANES

An important category of membranes is represented by nonporous (tight) polymeric membranes, because they are applied in many membrane separation processes. In all these processes the permeants are sorbed into the membrane material under the influence of their thermodynamic potential and pass it as a result of a driving force exerted.

The various processes are distinguished by the applied driving forces and by the phase condition(s) of the permeants. Let us consider these processes.

In pervaporation the feed is in the liquid phase; the permeants are evaporated by evacuation. The driving force is the activity gradient of the individual permeants across the membrane, which corresponds to the gradient in partial vapor pressure of those components.

In vapor permeation the driving force is also the gradient in partial vapor pressure. However, the feed is supplied as a vapor. Therefore, no heat supply is needed for evaporating the permeants as is required in pervaporation.

Membrane separation of gas mixtures resembles vapor permeation, the driving force being the gradient of the partial pressure of the gases involved.

With reverse osmosis the driving force is also a pressure gradient, but the feed (retentate) as well as the permeate are liquids. In this process the permeants are transported in the direction of increasing concentration. As a consequence, the osmotic pressure of the individual permeants must be overcome, which limits the application of reverse osmosis to dilute solutions of the retained compounds.

Also with dialysis (osmosis) the feed is a liquid mixture or solution. The permeants are diluted at the permeate side by means of a so-called sweeping solvent. Hence, in dialysis the driving force is a concentration gradient. If the boiling point of the permeants is much lower than that of the sweeping liquid, so that they can be separated by flashing and/or stripping from the latter, the dialysis process is called pertraction.

2.4. SELECTION OF MEMBRANE SEPARATION PROCESSES AND MECHANISMS

Although the review of the membrane processes and separation mechanisms in the previous sections is far from complete, it shows that the possibilities of the membrane separation technique are almost unlimited.

The mechanisms used in the processes of Section 2.1 and in the liquid membranes (especially with mediated transport, see Section 2.2) are all different and are very specific for those processes; for more information the reader is referred to the relevant literature (see also the references).

In this book we restrict ourselves to transport in tight (polymeric membranes), which covers the majority of those processes in which separation takes place on a molecular scale. It will be shown that all these processes can be described by the same physical model.

3

SURVEY OF MEMBRANE SEPARATION MODELS

For the description of membrane permeation, many physical models have been proposed and are mainly based on thermodynamics and/or statistical mechanics, thus expressing permeation rates by parameters derived from bulk properties. Several authors relate the sorption and transport phenomena to molecular properties such as rigidity of polymer backbones, types of pendant group, molecular packing density, and interactions with polar groups,[20, 21] but this is all still very qualitative. Of course, the latter route represents the most fundamental scientific approach. However, calculating molecular dynamics is still a long way off. Up to the present the thermodynamic/statistical approach is the most practical procedure.

In the existing theories a distinction is made between the membrane separations of liquids and that of gases. For liquid separations different theories have been developed. In principle, they can be placed in three main categories:

Irreversible thermodynamics,
Preferential sorption–capillary flow theory,
Solution-diffusion theory.

For gas separations (with tight membranes) only the solution-diffusion theory is used.

A comprehensive review of these models has been published by Soltanieh and Gill[2] and will be summarized briefly below.

3.1. IRREVERSIBLE THERMODYNAMICS

In models based on nonequilibrium or irreversible thermodynamics (IT) the membrane is treated as a "black box" barrier separating two phases far from equilibrium. This procedure is useful, especially when the structure of the membrane is not known and the mechanism of transport within it is not fully understood. Less information is required to set up the model, but less information can be obtained from it about the transport mechanisms inside the membrane.

For application of irreversible thermodynamics the basic assumption is made that the system can be divided into small subsystems in which "local equilibrium" exists, so that they can be described by thermodynamic parameters.[22] Thus at least locally the system should not be too far from equilibrium, which might be correct for the slow transport processes observed in membranes. According to Prigogine[23] this is about true when the transport processes are occurring in the range of linear-rate laws.

During a spontaneous irreversible process the entropy increases and, as a consequence, free energy is dissipated. The dissipation rate per unit volume is denoted by Φ, which according to Katchalsky and Curran[24] can be expressed by the general equation

$$\Phi = \sum_{i=1}^{n} J_i F_i \tag{3.1}$$

if J_i and F_i are the generalized fluxes and driving forces, respectively.

In IT Eq. (3.1) is applied in combination with the phenomenological equation postulated by Onsager[25]:

$$J_i = \sum_{k=1}^{n} L_{ik} F_k \qquad (i = 1, 2, 3, \dots, n) \tag{3.2}$$

which should be valid for sufficiently slow processes. This "linear law" states that in a system of n simultaneous flows, any flow J_i is linearly proportional to its "conjugate" force (the proportionality coefficient being expressed by the "straight coefficient" L_{ii}) and to the "nonconjugated" forces (the proportionality coefficients being denoted by the "cross coefficients" L_{ik}).

Substitution of Eq. (3.2) in Eq. (3.1) yields

$$\Phi = \sum_{i=1}^{n} \left(\sum_{k=1}^{n} L_{ik} F_k \right) F_i > 0 \tag{3.3}$$

since thermodynamics requires that in irreversible processes the entropy increases.

It can be shown[25] that in order to satisfy Eq. (3.3) the following two conditions must be fulfilled:

$$L_{ii} \geqslant 0$$

$$L_{ii} L_{kk} \geqslant L_{ik}^2 \tag{3.4}$$

A disadvantage of Eq. (3.3) is that, in systems with several fluxes and forces, the number of phenomenological coefficients becomes too large for experimental determination. Onsager, however, derived that for "near equilibrium" processes the matrix of the cross coefficients is symmetrical, hence

$$L_{ik} = L_{ki} \tag{3.5}$$

This "Onsager reciprocal relation" (ORR) reduces the number of phenomenological coefficients considerably. For instance, by applying ORR, the number of coefficients in a two-component system is reduced from 4 to 3 and with three components from 9 to 6.

An extended description of the thermodynamics of irreversible processes and of the Onsager relations is given in Appendix I.

The first IT-based membrane model was developed by Kedem and Katchalsky.[26] For isothermal, aqueous nonelectrolyte solutions of a single solute the following equations are obtained:

$$J_V = L_V(\Delta P - \sigma\, \Delta\pi_w)$$

$$J_s = C_{s,av}(1-\sigma)J_V + C_{s,av}\omega\, \Delta\pi_w \tag{3.6}$$

in which J_V is the total volume flux, J_s the molar solute flux, L_V the filtration coefficient, σ the Staverman reflection coefficient (coupling coefficient), ΔP the permeation pressure, $\Delta\pi_w$ the osmotic pressure difference of water across the membrane, $C_{s,av} = \Delta C_s/\Delta \ln C_s$ is the logarithmic mean of the solute concentrations across the membrane, and ω is the solute permeability at zero volume flux. The Kedem–Katchalsky equations are derived in Appendix I, Section (I.7).

In this derivation the membrane is treated as a barrier between the retentate and permeate phase and the gradients have been replaced by difference quantities across the membrane. If the concentration profile within the membrane is not linear (which is the case already when moderate swelling of the membrane material in the permeants occurs), it may be expected that the resulting phenomenological coefficients L_V, σ, and ω are concentration- and composition-dependent. This has indeed been observed experimentally.[2]

Many investigators examined the validity of the Onsager relations and the ORR experimentally, and observed considerable deviations especially in cases of high fluxes and large concentration gradients. Owing to the limited range of validity of these "linear laws" Spiegler and Kedem[27] derived equations for local transport, resulting more or less in a differential form of Eqs. (3.6). More constant phenomenological coefficients were obtained with these equations.

For more information about models based on IT the reader is referred to Appendix I and to the original paper of Soltanieh and Gill.[2]

3.2. PREFERENTIAL SORPTION–CAPILLARY FLOW THEORY

As pointed out by Soltanieh and Gill,[2] the concept of simple "sieve filtration" for separation on the basis of molecular size is ruled out in reverse osmosis, because for solutions such as sodium chloride–water a highly selective separation is observed, although the sizes of salt and water molecules are almost the same.

Reid and Breton[28] proposed a model in which the membrane is preferentially wetted by water, forming an absorbed film that prevents solute ions from entering the membrane material.

The same principle is used, e.g., by Sourirajan[1,29] who furthermore assumes the membrane material to be highly porous and heterogeneous, so that the mechanism of reverse osmosis is partly governed by surface phenomena and partly by fluid transport under pressure through capillary pores. Polymeric membranes with low dielectric constant, such as cellulose acetate, repel ions in the close vicinity of the surface, resulting in preferential sorption of water. The water layer thus formed is pressed through the pores, as shown schematically in Fig. 2. According to this figure there exists a critical pore size that yields optimum solute retention and fluid permeability, which should be twice the thickness of the sorbed water layer, t_w. According to Sourirajan, this pore size could be several times larger than salt or water molecular diameters and still show reasonable separation.

The model is based on a capillary flow model with viscous flow of the water and pore diffusion of the solute, while the "film theory" is applied for

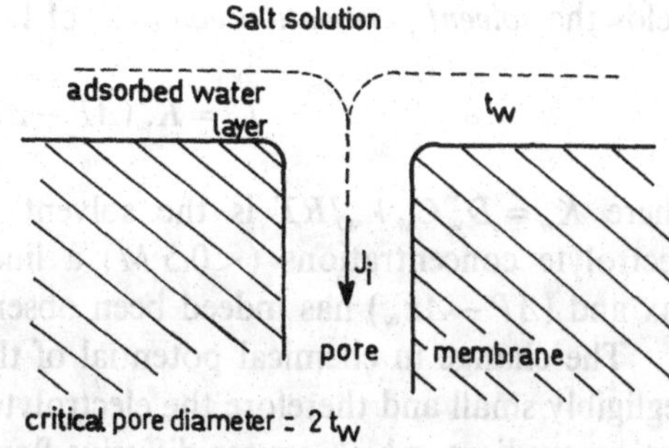

Figure 2. Preferential sorption capillary flow; model of Sourirajan.

salt transport through the adsorbed water layer. The basic equations correspond to those of the solution-diffusion model, which is discussed below, except that the concentration at the membrane surface on the feed side has been related to the bulk concentration via this film theory.[2]

3.3. THE SOLUTION-DIFFUSION MODEL

According to the solution-diffusion model, developed by Lonsdale *et al.*,[30] each permeant dissolves in the membrane material and passes by diffusion in response to its gradient in the chemical potential. For aqueous electrolyte solutions he derived that the water flux J_w is expressed by

$$J_w = \frac{D_w C_w}{RT} \frac{d\mu_w}{dz} \approx \frac{D_w C_w}{RT} \frac{\Delta\mu_w}{l} \tag{3.7}$$

where D_w is the diffusivity of water in the membrane, C_w the concentration of water in the membrane, $d\mu_w/dz$ the gradient of the chemical potential of water, R the gas constant, T the temperature (Kelvin), and l is the membrane thickness. In this equation it is tacitly assumed that D_w and C_w in the membrane are constant.

For isothermal systems

$$\Delta\mu_w = RT\,\Delta \ln a_w + V_w\,\Delta P \tag{3.8}$$

if a_w is activity and V_w the partial molar volume of water, respectively.

Substituting in Eq. (3.7) and using Lewis's equation for the osmotic pressure (π)

$$V_w\,\Delta\pi_w = -RT\,\Delta \ln a_w \tag{3.9}$$

yields the *solvent permeation equation* of Lonsdale:

$$J_w = K_w(\Delta P - \Delta \pi_w)/l \tag{3.10}$$

where $K_w = D_w C_w V_w/RT$ is the solvent permeability. At not too high electrolyte concentrations ($<0.5\ M$) a linear relationship between water flux and $(\Delta P - \Delta \pi_w)$ has indeed been observed experimentally.[2]

The change in chemical potential of the electrolyte due to pressure is negligibly small and therefore the electrolyte flux is only due to the concentration gradient, which causes diffusive flow.

By assuming equal distribution coefficients of the solute, k_s at the upstream and downstream sides of the membrane and applying Fick's law, Lonsdale obtained the following equation for the *solute flux*:

$$J_s = D_s k_s \frac{(C_R - C_P)}{l} \tag{3.11}$$

where C_R and C_P are the solute concentrations in the retentate and permeate, respectively, and D_s is the solute diffusivity in the membrane material.

According to Eq. (3.11) the solute flux is independent of permeation pressure, while the solvent flux increases proportionally to it [see Eq. (3.10)]. Therefore, the selectivity must increase with pressure, which is indeed observed experimentally.[2]

Usually, also the *solute* flux increases somewhat with pressure; the latter is attributed to small leaks in the membranes, causing nonselective Poiseuille flow of the concentrate phase through the membrane. This flow hardly affects the much larger solvent flow, but it influences the solute flux considerably due to the extremely small diffusive flow of the latter. Therefore, many equations have been proposed that account for convective flow. An extensive survey of these models is given in the publication of Soltanieh and Gill.[2]

The Lonsdale equations are widely applied, in spite of the many simplifications made in their derivation that are responsible, e.g., for the fact that these equations are not capable of predicting coupling effects.

3.4. VISCOUS FLOW MODELS; ACCOUNTING FOR IMPERFECTIONS

Convection effects have been ignored in the derivation of the various membrane separation equations in the previous sections. It is expected, however, that many real membranes contain micropores through which convective–diffusion transport takes place under (reverse) osmosis conditions. If the bulk flow through a membrane is denoted by N and the total (diffusive + convective) flow of component i through that membrane by N_i, then

$$N_i = J_i + x_i N \tag{3.12}$$

if molar quantities are used.

In the case of dilute solutions, the first term on the right-hand side of this equation is predominant for the solvent and hence the solvent flux is hardly affected by convection (leakage). For the repelled solute, however, the nonselective second term prevails, especially at high intrinsic selectivity of the membrane material. Because of the irregular formation of the number and sizes of imperfections, the transport behavior of the solute is rather complicated and cannot be described by any equation from the previous sections. Therefore, the various transport models have been extended for convective flow according to Eq. (3.12).

In the solution-diffusion-imperfection model, which has been developed by Sherwood *et al.*,[31] an additional term has been included for pore flow, requiring a third parameter. This model shows a very close fit with the experimental data, but all three parameters in this relation vary with feed concentration and permeation pressure, thus limiting its practical application.[2]

In other models transport is assumed to take place exclusively through pores.[32–34] According to these models coupling occurs by viscous flow. Separation is achieved only if the solute concentration in the pores differs from that in the bulk feed, due to whatever mechanism may exist. The total volume flux J_V is described by Poiseuille flow:

$$J_V = \frac{\varepsilon r_p^2 \Delta P}{8\eta\tau l} \tag{3.13}$$

in which ε is the fraction open-pore area, η the viscosity of the fluid in the pores, τ the tortuosity, and r_p the equivalent pore radius.

The total solute flux equals the sum of a contribution for viscous flow and another due to diffusion with respect to that flow, as expressed by Eq. (3.12). Also, these models lead to equations with three parameters, which must be determined experimentally and generally exhibit a dependence on concentration and permeation pressure.

A model based on irreversible thermodynamics was proposed by Soltanieh and Gill.[2] They started from the Spiegler–Kedem differential equations,[27] extended for convection, and solved them without applying ORR. Again a three-parameter equation was obtained.

All models mentioned predict a nonlinear relationship between solute concentration and total volume flux according to the same type of equation:

$$C_R/C_P = K_1 - K_2 \exp(-K_3 J_V) \tag{3.14}$$

except that the physical meaning of the three parameters K_1, K_2, and K_3 differ.[2]

3.5. MODELS FOR THE SEPARATION OF GAS (VAPOR) MIXTURES

The transport models described in the previous sections concern the separation of aqueous solutions and liquid mixtures. Also, various permeation equations exist for membrane separation of gas mixtures and are all based on the solution-diffusion model.

One might expect that the models for gas separation are simpler than those for liquids. It turned out, however, that for a good description of gas separation equations had to be developed which are at least as complicated as those for liquid separation. This is mainly due to deviating sorption behavior of different gases and gas mixtures in the various polymers used as membrane materials.

In 1866 the first investigation of membrane separation of gas mixtures was presented by Thomas Graham. He observed the preferential permeation of oxygen from air through natural rubber membranes.

Wroblewski published in 1879 the first permeation rate equation for gases passing through rubber membranes in which he combined Henry's law for gas sorption with Fick's law for diffusion, thus introducing the solution-diffusion model[35]:

$$J_i = D_i k_i \,\Delta P/l \tag{3.15}$$

in which D_i and k_i represent the diffusivity and Henry coefficient of gas i, respectively. According to this equation the rate at which gases permeate through tight polymer membranes is proportional to the applied pressure gradient of those gases across the membranes, the proportionality being expressed by the permeability.

If a gas displays low solubility in the membrane material, the permeability is a constant. At high solubilities no linearity exists in general, while the selectivity of the gas separation is lower than that calculated from the pure gas permeabilities applying conventional permeation theories. These deviations are often attributed to specific interactions between the gases and the polymer.[20]

A distinction is made between rubbery polymers, being above their glass-transition temperature (T_g), and glassy polymers, being below that temperature. The thermal volumetric expansion coefficient of a polymer is smaller in the glassy state than in the rubbery state (see Figure 3). This is attributed to the extraordinarily long relaxation time for polymer chains below the glass-transition temperature, resulting in a nonequilibrium excess volume in glassy polymers.[20]

The gas sorption coefficient (Henry coefficient) of rubbery polymers is either constant or else increases with pressure, due to "plasticization" (or swelling) of the polymer in the permeants, enhancing the mobility of the polymer chains. In the latter case the permeability also increases with pressure. In glassy polymers the gas sorption coefficient usually decreases with pressure. This effect is explained by the "dual-mode" sorption model, which is based on the concept that microvoids are frozen into the matrix of glassy polymers.[36,37] Gas sorption takes place partly by absorption into the bulk polymer, according to the linear Henry's law, and partly by adsorption onto the surfaces of the cavities in the polymer, according to a Langmuir type of sorption process. This model results in the overall equation[38]

$$C_i = k_i P_i + \frac{C'_{Hi} b_i P_i}{(1 + b_i P_i)} \tag{3.16}$$

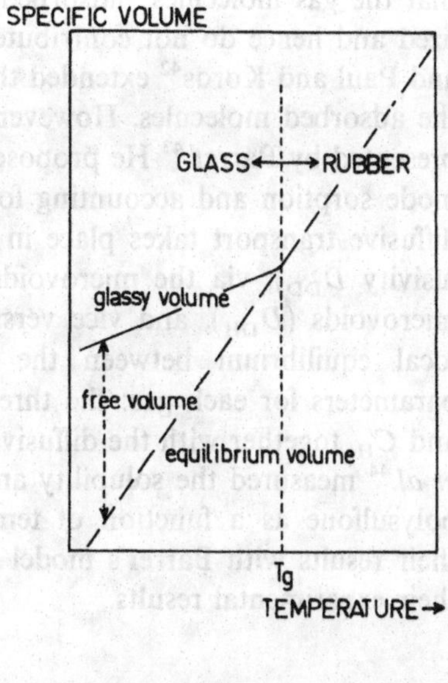

Figure 3. Glass transition of polymers.

in which C_i is the gas concentration in cm^3(STP) per cm^3 of polymer, P_i is the partial pressure of gas i in bar, k_i is the Henry coefficient in $cm^3(STP)/(cm^3 \cdot bar)$, C'_{Hi} is the Langmuir capacity constant in $cm^3(STP)/cm^3$, and b_i is the Langmuir affinity constant (bar^{-1}).

According to Sanders and Koros[39] it is fundamentally more correct to apply fugacity f_i instead of partial pressure P_i, which in our opinion is true. Especially in the case of condensable gases (like CO_2) and vapors near their saturation pressure, large deviations from the ideal gas law may occur. By applying fugacity the nonideal behavior of the gas (vapor) phase is taken into account.

For binary gas mixtures (of gases i and j) Eq. (3.16) can then be transformed into

$$C_i = k_i f_i + \frac{C'_{Hi} b_i f_i}{(1 + b_i f_i + b_j f_j)} \tag{3.17}$$

Many authors have shown that, in general, gas solubility in glassy polymers can be calculated accurately by means of these equations.[38,40] It must be kept in mind, however, that for each gas three parameters are used for fitting the experimental data. Notwithstanding the physical meaning attributed to these parameters, they must be derived from the sorption measurements themselves in order that a satisfactory description be achieved.

Vieth and Sladek[37] derived a permeation rate equation by combining the dual-mode sorption equation and Fick's diffusion equation, assuming that the gas molecules "adsorbed into the holes" are completely immobilized and hence do not contribute to the transport process. Petropoulos[41] and Paul and Koros[42] extended this model by assuming partial mobility of the adsorbed molecules. However, the most complete treatment has been presented by Barrer.[43] He proposed a mathematical model, based on dual-mode sorption and accounting for the fact that, according to this model, diffusive transport takes place in the bulk polymer (described by the diffusivity D_{DD}), via the microvoids (D_{HH}), from the bulk polymer to the microvoids (D_{DH}), and vice versa (D_{HD}). This equation, in which again local equilibrium between the two modes is assumed, contains six parameters for each gas: the three dual-mode sorption parameters k_i, b_i, and C'_{Hi} together with the diffusivities D_{DD}, D_{DH}, and $(D_{HH} + D_{HD})$. Sada *et al.*[44] measured the solubility and permeability of pure CO_2 and CH_4 in polysulfone as a function of temperature and pressure. They compared their results with Barrer's model and observed an excellent description of their experimental results.

3.6. CONCENTRATION POLARIZATION

In membrane separation processes there is a concentration gradient near the membrane interface due to the buildup of the retained compounds. This effect is called concentration polarization. It increases with increasing permeate flux, which reduces the driving force for permeation, resulting in a lower flux and a less selective separation.

Let us consider membrane separation of liquids first. Suppose J_i represents the flux of the retained component i through the membrane while J represents the total membrane flux (see Figure 4). At steady state and one-dimensional transport (in the z-direction) J equals also the overall convective flux in the direction of the membrane. From a simple mass balance, assuming Fick's diffusion law to be valid, it follows that

$$J_i = -D_i \frac{dC_i}{dz} + C_i J \tag{3.18}$$

The first term on the right-hand side of this equation represents the back-diffusion of component i into the bulk liquid phase, and the second term the total convective flux of i toward the membrane. Because J_i/J is the concentration C_{pi} of i in the permeate, which is constant at steady state, Eq. (3.18) can be integrated between the boundaries

$z=0$, $C_i = C_{mi}$ = concentration of i at the interface of the membrane and the retentate

and

$z=-\delta$, $C_i = C_{bi}$ = concentration of i in the bulk retentate phase

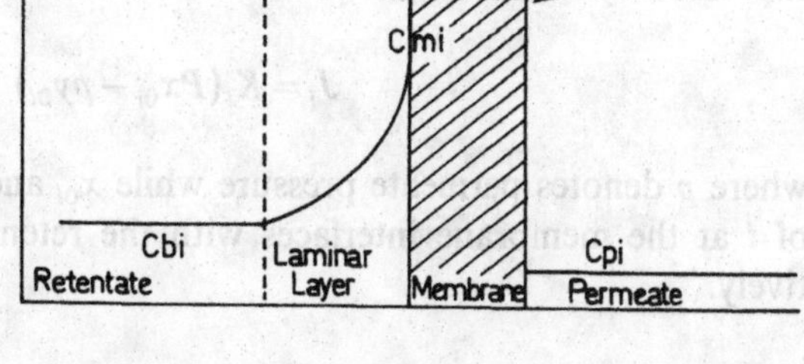

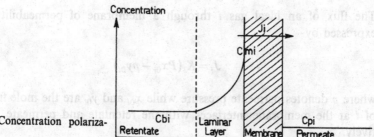

Figure 4. Concentration polarization.

(where δ is the laminar diffusion layer) to yield

$$\frac{(C_{mi}-C_{pi})}{(C_{bi}-C_{pi})}=\exp(J/k_i) \tag{3.19}$$

in which $k_i=D_i/\delta$ is the mass transfer coefficient. Correspondingly, such an equation can be derived for eventual concentration polarization effects at the permeate side of the membrane.

According to Eq. (3.19) the concentration of the retained component at the membrane interface increases with respect to its concentration in the bulk phase if the ratio of the total flux and the mass transfer coefficient increases, thus reducing the concentration gradient across the membrane of the component to be separated, and hence its flux. This unfavorable effect of concentration polarization can be reduced by increasing the mass transfer coefficient. The latter can be achieved by maintaining a high flow rate of the liquid phase along the membrane surface and by applying turbulence promoters (spacers) between the membranes.

For gases it is convenient in Eq. (3.18) to express the concentrations in mole fractions. If, furthermore, we express the fluxes in volume rates at standard pressure P_0, this equation transforms into

$$J_i=-\frac{P}{P_0}D\frac{dx_i}{dz}+x_iJ \tag{3.20}$$

if P is the feed (retentate) pressure and D the mutual diffusivity of the gas mixture.

According to the kinetic gas theory of ideal gas mixtures, D is independent of concentration and inversely proportional to the pressure. Therefore

$$\frac{P}{P_0}D=D_0=\text{constant} \tag{3.21}$$

The flux of an ideal gas i through a membrane of permeability K_i is expressed by

$$J_i=K_i(Px_{0i}-py_{0i}) \tag{3.22}$$

where p denotes permeate pressure while x_{0i} and y_{0i} are the mole fractions of i at the membrane interfaces with the retentate and permeate, respectively.

The following equation can be derived[45] from Eqs. (3.20)–(3.22) (see Appendix II):

$$x_i = Y_i\left[1+\left(\frac{J}{K_i(P-p)}-1\right)\left(1-\frac{p}{P}\right)\exp\left(-\frac{J}{b_1}\right)\right] - \frac{p}{P}(Y_i - y_i)\exp\left(-\frac{J}{b_1}-\frac{J}{b_2}\right) \qquad (3.23)$$

where x_i (y_i) is the concentration of i in the bulk retentate (permeate) phase, Y_i is the local permeate concentration of i, and b_1 $(b_2) = D_0/\delta_1$ (D_0/δ_2) is the backdiffusion factor at the retentate (permeate) side.

Equation (3.23) relates the local permeate concentration to the bulk permeate and retentate concentration, depending on permeation conditions and membrane properties. It predicts optimum separation if the backdiffusion factor (which in essence represents a kind of mass transfer coefficient) $b \gg J$, while the selectivity ceases if $b \ll J$.

The second term on the right-hand side of Eq. (3.23) represents a contribution to the mode of membrane operation. For countercurrent flow of permeate and retentate it predicts increasing selectivity, and for cocurrent flow reduced selectivity. Calculations show that these effects are maximum for low-selectivity membranes combined with a relatively high backdiffusion factor. For cross flow, which in fact corresponds to the conditions where $b_2 \ll J$, the second term vanishes. The latter mode of operation occurs in all presently available gas modules, due to the presence of a porous support, which causes significant mass-transfer limitations.

In cases of highly selective membranes (pure gas permeability ratio >50) no effect of the different modes of membrane operation (on separation) is predicted. However, a tremendous effect of the backdiffusion factor on the retentate side on flux is found (see Figure 5). A corresponding reduction in selectivity is calculated.

Experiments with different mixtures of H_2 and CO_2 showed that the backdiffusion factor correlates well with the square root of the normal volume flow rate of the gas along the membrane surface, and is almost independent of composition and total pressure.[45] Its magnitude proves to be such that for permeabilities of nonhydrogen-containing gas mixtures above about $10\ m^3(STP)/(m^2 \cdot bar \cdot day)$ concentration polarization will drastically reduce flux and selectivity for the conditions envisaged. In the existing commercial membrane gas separation units the fluxes are relatively low $[<1\ m^3(STP)/(m^2 \cdot day)]$, so that effects of concentration polarization are only minor. However, for economical reasons highly permeable gas membranes are here being developed. It is shown that there is no point

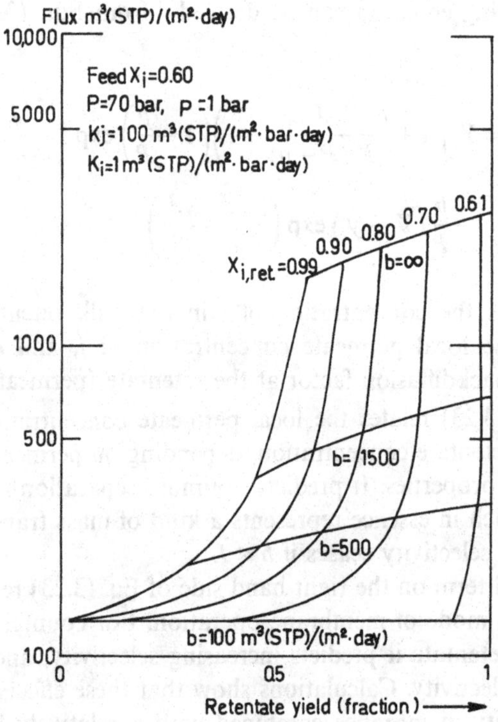

Figure 5. Effect of backdiffusion factor on flux.

in increasing the flux considerably without considering the design of the membrane module and the planned operating conditions, because then concentration polarization may severely hamper the potential performance of such a high-flux membrane.

3.7. BLOCKING, FOULING, AND POISONING

A large disadvantage of microporous and ultrafiltration membranes is their liability to pore blocking, which may lead to a dramatic reduction in flux thus hampering their practical application. Mechanical pore blockage takes place if particles are present of sizes about equal to or somewhat smaller than the effective pore openings in the membrane. Blockage can occur also by dissolved species. Due to, e.g., concentration polarization the solubility of the retained compounds can be exceeded causing precipitation by crystallization inside the pores. This effect has been observed in the concentration of whey by means of ultrafiltration membranes. Precipitation of calcium phosphate took place mainly inside the membrane.[46]

In fact no real solution exists for this problem. By proper choice of the type of membrane, suitable operating conditions, and application of intermediate backflushings, reasonable fluxes can still be maintained, although they are much lower than the pure liquid fluxes.

Fouling of tight (nonporous) membranes also occurs in practice. If this is due to physical or physicochemical interactions, it can be reduced by promoting turbulent flow along the membrane surface and applying intermediate (chemical) cleaning. In aqueous solutions or suspensions, less fouling is observed with strongly hydrophilic membranes (e.g., from cellulose acetate or ceramics) due to the preferential wetting of such materials by water. For the same reason hydrophobic membranes (e.g., from polyolefins) are less fouled by solutions or suspensions in nonpolar organic solvents.

As a result of the relatively low fluxes obtained with tight membranes the effect of physical and physicochemical fouling on the flux is only minor, contrary to what is generally expected. If, however, strong (chemical) interactions take place between some species from the solution and the membrane surface, the properties of the membrane can be dramatically and irreversibly changed. A well-known effect is the preferential adsorption of hydroxycarbonic acids even from very diluted aqueous solutions onto cellulose acetate membranes. The water flux decreases strongly with increasing amounts of adsorbed solute, while the selectivity of the membrane for the solute increases with respect to that of water. This type of fouling is called poisoning and cannot be prevented or reduced. In such cases a different membrane material must be chosen.

This brief summary about fouling has been presented for the sake of completeness and because it plays such an important role in the practical application of membrane processes. A more extended treatment of the subject lies beyond the scope of this book. A comprehensive survey of fouling and its treatment has been given elsewhere by Bootsveld and Wienk.[47]

4

COMPARISON OF MEMBRANE PERMEATION MODELS

4.1. LIQUID SEPARATIONS

The main purpose of a membrane permeation equation is the reliable prediction of the behavior of membrane separation systems. An optimum permeation model preferably uses no system-related properties, but rather physical and thermodynamic data, because then a larger predictive capability is achieved with minimum experimental effort. In the IT-based equations the phenomenological coefficients, which are system-related, must be determined experimentally. These coefficients were found to be dependent on concentration and composition, probably due to the fact that the Onsager relations, applied in the derivation of the IT-membrane equations, are not valid. As a consequence, it is problematic to find empirical correlations for the prediction of the coefficients and, if so, then only at the expense of laborious effort.

The preferential sorption–capillary flow (PSCF) model and the solution-diffusion model assume transport mechanisms and, accordingly, fluxes are related to forces acting in those systems. Consequently kinetic, physical, and thermodynamic properties of the membranes and permeants are involved in these transport models. If the required information is available, prediction of membrane performance is possible without experimental data under actual operating conditions. Therefore these models are potentially advantageous compared to the IT equations.

In the case of a single solute the PSCF model of Sourirajan, extended by Dicksen *et al.*[48] for boundary layer transfer, contains four parameters: the pure water permeability, the solute transport parameter, the solute concentration on the membrane surface at the feed side, and the solute mass transfer coefficient (which contains the thickness of the adsorbed water layer). In practice, the latter two parameters must be determined empirically and depend on the membrane properties and on operating conditions, impairing its advantage in comparison to the IT models.

Another objection to the PSCF model is the assumed presence of "molecular size" pores in the membrane, through which solvent transport takes place via viscous flow. The existence of such pores cannot be shown in a direct manner, but is derived indirectly from the PSCF model itself. In any case, it is not a general model, because we have shown[49] that separation of hydrocarbon mixtures at high fluxes and selectivities is possible with elastomeric membranes. Such materials are highly swollen in the permeants and certainly do not contain pores; they must be regarded as homogeneous solutions of a polymer and permeants in accordance with the Flory–Huggins theory.[50,51]

A more general description is represented by the solution-diffusion model.

For a *single solute* system the Lonsdale equations contain only two parameters: the solute permeability and the solvent permeability; see Eqs. (3.10) and (3.11). These equations are derived for uncoupled transport and hence are incapable of predicting mutual effects on simultaneous transport of various permeants as observed experimentally.

Several authors have extended the solution-diffusion model for convection, in view of the supposed presence of imperfections (pores) in the membranes. Accordingly, coupling takes place by (nonselective) leakage of the permeate.[2] In this way the number of parameters is increased; at least part of them must be determined experimentally, reducing the predictive capability of the model. Corresponding objections hold for the viscous (convective) flow models.

In our opinion the restricted validity of the Lonsdale equations as well as that of the other models considered is caused by the simplification of assuming in their derivation a linear concentration gradient inside the membrane, which contradicts experimental observations of Tock *et al.*[52] and Mulder *et al.*,[53] who showed almost exponential gradients. The diffusivity is usually tremendously affected by concentration.[54] The local permeability corresponds to the product of concentration and diffusivity. Therefore, large deviations can be expected from the assumption of a linear concentration gradient. It is mainly this assumption that eliminates coupling effects, and therefore a reliable model can be achieved only without this simplification.

4.2. GAS SEPARATIONS

The gas permeability equations commonly used are based on the solution-diffusion model, in which the solubility is expressed by the dual-mode sorption model and the diffusion rate by Fick's law. Although various authors report excellent description of experimental results by this model, it must be emphasized that a large number of (fitting) parameters are available, and this enables perfect matching.

Physically, the existence of the proposed dual-mode sorption process is very unlikely. The supposed microvoids in a polymer could never be made visible. This means that, considering the ultimate resolution of the available techniques, the size of the voids must be less than 2 nm, which is thermodynamically very unstable. In our opinion another mechanism (viz. the local orientation of polymer molecules) is responsible for the behavior of the "glassy" polymers (see Section 6.5).

4.3. A NEW PERMEATION MODEL FOR GASES AND LIQUIDS

In this study a new model is derived, based on the solution-diffusion theory, which is valid for gases as well as for liquid mixtures. It accounts for the various shortcomings in the conventional equations and is capable of predicting the large effects of crystallinity and swelling (solubility) of the membrane material on permeability.[4]

The study includes a modified Maxwell–Stefan equation for calculating diffusive transport, a modified Flory–Huggins equation to calculate the solubility of the permeants in the membrane material, and a modified Vignes equation for calculating the mixture diffusivities. Procedures are given for determining or estimating the required kinetic, physical, and thermodynamic properties.

5

BASIC DIFFUSION EQUATION

5.1. INTRODUCTION

Diffusion is a process by which matter is transported as a result of random molecular motion. If in a system a substance is not distributed homogeneously, a net transfer of that substance takes place in the direction of lower concentration. Fick stated in 1855 that in isotropic systems the rate of transfer (J_i) of diffusing matter i through a unit area of a section is proportional to the concentration gradient ($\delta C_i/\delta z$) measured normal to that section, namely,

$$J_i = -D_i \, \delta C_i/\delta z \tag{5.1}$$

in which D_i is Fick's diffusivity. The diffusivity defined according to Eq. (5.1) appears to vary strongly with composition and concentration especially in liquid mixtures. Modern diffusion theories[54] have adopted the premise that the diffusive flow is proportional to the gradient in the *chemical potential* ($\delta\mu_i/\delta z$). This theory leads to an activity-corrected diffusion coefficient ($\mathbf{D_i}$):

$$\mathbf{D_i} = \frac{D_i}{[\delta \ln a_i/\delta \ln x_i]_{T,P}} \tag{5.2}$$

McCall and Douglass[55] showed experimentally that the Fickian diffusivity is indeed almost proportional to $\delta \ln a_i/\delta \ln x_i$ in various binary mixtures (see Figure 6), but although $\mathbf{D_i}$ is less concentration-dependent than D_i, it is still not a constant.

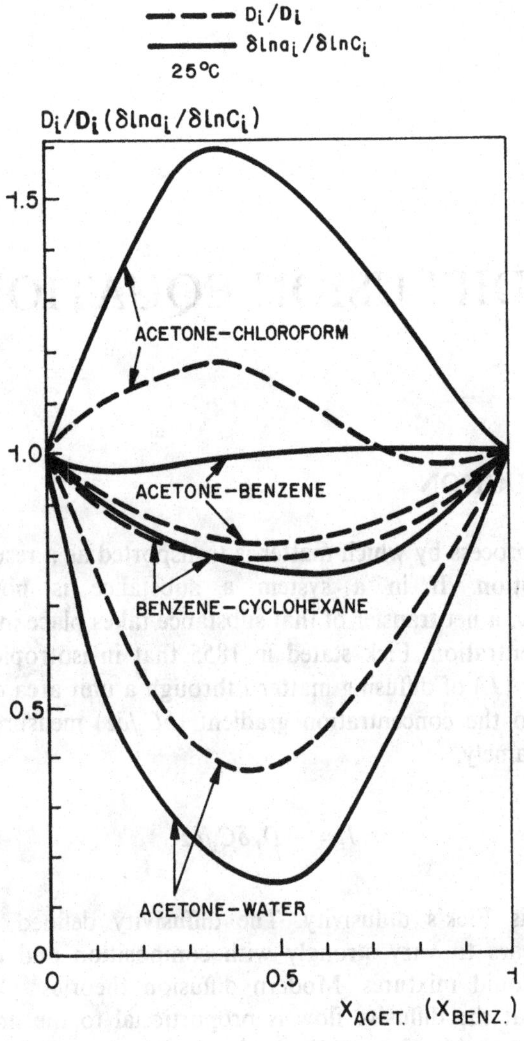

Figure 6. Effect of activity on diffusivity.

Another phenomenon, which has been widely recognized, is that in systems with components of different diffusivities the resulting net diffusive mass transfer is in general compensated by a convective bulk flow of the mixture in order to maintain the requirements of an isobaric, isothermal system. If the total molar mass transfer of matter i is denoted by N_i and the bulk flow by N, the diffusive flux of component i is expressed by

$$J_i = N_i - x_i N \tag{5.3}$$

5.2. THE MAXWELL–STEFAN EQUATION

It is common practice to apply the "frictional model of transport" for deriving a diffusion equation, that accounts for convective flow. According to this model the driving force for transport (expressed by the gradient in the chemical potential, $-\text{grad}\ \mu_i$) exerted on a species i in a mixture is, in steady flow, balanced by the frictional forces exerted on it by the other components present in that system. On the assumption that the average friction exerted on molecules i in a multicomponent system equals the molar average of the frictions experienced by i in binary mixtures of i and the individual components of the former mixture, the well-known Maxwell–Stefan equation has been derived (see Appendix III):

$$C_i \frac{\text{grad}\ \mu_i}{RT} \sum_{\substack{j=1 \\ j \neq i}}^{n} \frac{x_i N_j - x_j N_i}{D_{ij}} \tag{5.4}$$

in which D_{ij} is the binary Maxwell–Stefan (MS) diffusivity of i in a mixture of i and j.

In order to derive this equation it was tacitly assumed that D_{ij} is independent of concentration and of the presence of other components (see Appendix III). In the case of gases, for which the diffusivities usually show only minor differences, Eq. (5.4) can be applied successfully. For mixtures of liquids[56] and of polymers and liquids,[4] however, it has been observed experimentally that D_{ij} varies considerably with composition and concentration, although much less than Fickian.

To summarize, we conclude that a more fundamental diffusivity equation is required for reliable prediction of mass transport in multicomponent mixtures.

5.3. EQUATION OF DARKEN, PRAGER, AND CRANK

Darken and Prager have related the diffusion coefficient in a binary mixture, measured by an experiment with a radioactive tracer, to the self-diffusion coefficient (also referred to as intrinsic diffusivity) in terms of thermodynamic properties of the system. We apply their procedure, as described by Crank,[57] to determine the diffusivity in a multicomponent system.

Let us consider a system comprising molecules i in a multicomponent mixture (m) and let the concentration C_i of i be maintained in an equilibrium condition by a force f_i per mole of i in the direction of the concentration gradient of i. The generalized form of the condition for this thermodynamic equilibrium is

$$f_i = \text{grad}\ \mu_i \tag{5.5}$$

The rate of transfer of i due to a force f_i is

$$J_i = -\frac{f_i C_i}{\sigma_i \eta_{\text{m}}} = -\frac{C_i}{\sigma_i \eta_{\text{m}}}\ \text{grad}\ \mu_i \tag{5.6}$$

where $\sigma_i \eta_{\text{m}}$ represents a friction coefficient, consisting of a molecular shape factor of i (σ_i) and the viscosity of the mixture (η_{m}).

But in the equilibrium condition, Eq. (5.6) is also the rate of transfer by diffusion relative to a section through which there is no bulk flow, and so we have

$$D_{i\text{m}}\ \text{grad}\ C_i = \frac{C_i}{\sigma_i \eta_{\text{m}}}\ \text{grad}\ \mu_i \tag{5.7}$$

We apply the same treatment to a multicomponent system of uniform chemical composition in which one of the components is partly labeled. The labeled molecules have a concentration gradient which is opposite and equal to that of the corresponding nonlabeled molecules. For this system we get

$$^*D_{i\text{m}}\ \text{grad}\ ^*C_i = \frac{^*C_i}{^*\sigma_i \eta_{\text{m}}}\ \text{grad}\ ^*\mu_i \tag{5.8}$$

where asterisks denote labeled molecules, $^*D_{i\text{m}}$ representing the coefficient of self-diffusion of i in the mixture (m).

Let us consider isobaric, isothermal conditions. Then, because of the ideality of the system,

$$\text{grad}\ ^*\mu_i = RT\ \text{grad}\ \ln\ ^*C_i$$

which, combined with Eq. (5.8), yields

$$^*D_{i\text{m}} = \frac{RT}{^*\sigma_i \eta_{\text{m}}} \tag{5.9}$$

Putting $^*\sigma_i = \sigma_i$ and substituting Eq. (5.9) into Eq. (5.6) yields the Darken, Prager, and Crank equation:

$$J_i = \frac{^*D_{im}}{RT} C_i \operatorname{grad} \mu_i \tag{5.10}$$

It is interesting to note that, according to Eq. (5.10), the coefficient of self-diffusion of component i in the mixture must be applied for calculating the diffusive flux of i.

We note that this equation has been derived for the condition when no convective flow occurs. In the next section we account for convective flow.

5.4. THE MODIFIED MAXWELL–STEFAN EQUATION

In the derivation of the MS equation (see Appendix III) it was assumed that the interaction between molecules i and j is not affected by the presence of other molecules, hence their mutual friction coefficient is independent of composition and concentration. This implies that, during a collision between a molecule i and a molecule j, no other molecules are involved. This might be about valid in gas mixtures, but it is certainly not true in the case of liquids. This is illustrated in Figure 7.

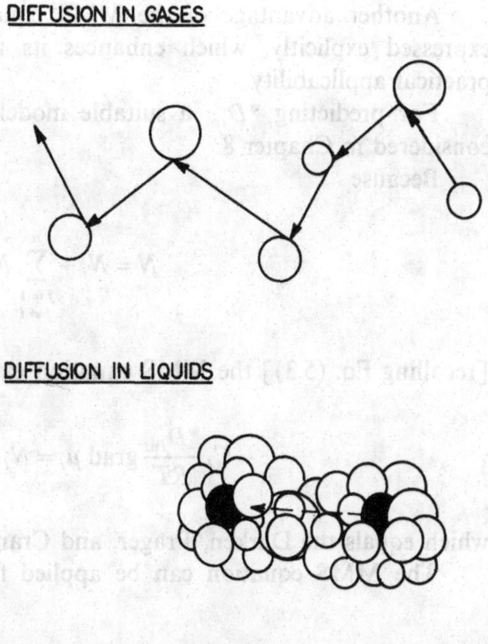

Figure 7. Interaction of molecules in gases and liquids.

In liquids each molecule is surrounded by a relatively large number of "direct neighbor molecules," expressed by the coordination number, being about 12 for spherical molecules. Therefore, it may be expected that a molecule i, during its movement in a liquid mixture, will experience a frictional force, which everywhere results from the combined interactions of all locally surrounding molecules. Accordingly, the friction coefficient will depend on size and shape of molecule i (σ_i) and on an average frictional property of the local mixture (η_{m}). Based on this principle the next equation is derived in Appendix IV:

$$C_i \frac{\operatorname{grad} \mu_i}{\sigma_i \eta_{\mathrm{m}}} = \sum_{\substack{j=1 \\ j \neq i}}^{n} (x_i N_j - x_j N_i)$$

which, after eliminating the friction coefficient by means of Eq. (5.9), yields the modified Maxwell–Stefan (MMS) equation:

$$\boxed{C_i \frac{{}^*D_{im}}{RT} \operatorname{grad} \mu_i = \sum_{\substack{j=1 \\ j \neq i}}^{n} (x_i N_j - x_j N_i)} \tag{5.11}$$

In the derivation of this equation no arbitrary assumption for the average friction has been introduced, as was done in the derivation of the MS Eq. (5.4), and therefore the former has a more general validity.

Another advantage of the MMS equation is that the diffusivity is expressed explicitly, which enhances its accessibility and facilitates its practical applicability.

For predicting ${}^*D_{im}$ a suitable model is still required. This will be considered in Chapter 8.

Because

$$N = N_i + \sum_{\substack{j=1 \\ j \neq i}}^{n} N_j$$

[recalling Eq. (5.3)] the MMS equation can be transformed into

$$-C_i \frac{{}^*D_{im}}{RT} \operatorname{grad} \mu_i = N_i - x_i N = J_i$$

which equals the Darken, Prager, and Crank equation.

The MMS equation can be applied for deriving a permeation rate equation for transport through tight polymer membranes. Considering perfect (nonleaking) membranes at steady-state conditions no convective flow is involved, so that this equation transfers into that of Darken, Prager, and Crank [Eq. (5.10)]. Both equations show that, apart from the driving force, the flux depends on the concentration (hence solubility) and the diffusivity of the permeants in the membrane material. In the following sections these parameters are considered individually.

6

SOLUBILITY OF PERMEANTS IN SEMICRYSTALLINE AND CROSSLINKED POLYMERS

6.1. SOLUBILITY OF LIQUIDS IN POLYMERS

The selectivity of, and flux in, tight polymer membranes are strongly affected by the solubility of the permeating compounds in the membrane material. Prediction of solubility requires a reliable theoretical equation, for which the Flory–Huggins[50,51] equation is usually applied. This equation relates the entropy of mixing to polymer size and composition and is derived by calculating the number of possible configurations by filling the sites of an imaginary lattice with constant coordination number Z successively with polymer molecules occupying m sites each and solvent molecules occupying single sites.

The authors further assumed that a polymer segment is sufficiently flexible to occupy each of the Z sites around its neighbor.

During the process of filling the lattice, the chance that a certain site is not yet occupied decreases with the number of molecules already placed. Flory assumed that this chance equals the fraction of unoccupied sites and arrived at an equation for the partial molar entropy of mixing for the solvent (ΔS_s^{Mix}):

$$\Delta S_s^{\text{Mix}} = -R[\ln \Phi_s + (1 - 1/m)\Phi_p] \tag{6.1}$$

where Φ_s (Φ_p) is the volume fraction of solvent (polymer), m the molar volume ratio of polymer and solvent, and R is the gas constant.

In a more fundamental approach Huggins accounted for the fact that, if a site is occupied by a segment of a previously placed polymer chain, then at least one of the Z surrounding sites is also occupied if that segment is the first or last one of that chain (see Fig. 39); at least two sites will be occupied if that segment is an intermediate one (see Fig. 38). In this way he obtained the equation

$$\Delta S_s^{\text{Mix}} = -R\left[\ln \Phi_s - \frac{Z'}{2}\left(1-\frac{1}{m}\right)\ln\left(1-\frac{2}{Z'}\Phi_p\right)\right] \tag{6.2}$$

where

$$Z' = \frac{Zm}{(m-1)(1-f_0)} - \frac{2f_0}{(1-f_0)}$$

f_0 being the (small) chance that a site is occupied by a segment of the chain being placed.

Huggins proposed expanding the logarithmic term of Eq. (6.2) in a series and ignoring terms in Φ_p^3 and higher. This approach leads to

$$\Delta S_s^{\text{Mix}} = -R\left[\ln \Phi_s + (1-1/m)\Phi_p + \frac{1}{Z'}(1-1/m)\Phi_p^2\right] \tag{6.3}$$

which, apart from the last term, is the same as Eq. (6.1).

In the statistical treatment applied by Flory and Huggins no preferential orientation is assumed, so their equation is in fact only valid for athermal absorption. In general, there is an enthalpy effect, however. The authors assumed that this effect, being relatively small, has only a minor influence on the entropy of mixing. Therefore, they derived the partial molar free energy of mixing ΔF_s^{Mix} simply by adding the enthalpy contribution to the partial molar mixing entropy, using for the former the solubility parameter model of Scatchard and Hildebrand[58]:

$$\Delta H_s^{\text{Mix}} = V_s(\delta_s - \delta_p)^2\, \Phi_p^2 \tag{6.4}$$

where ΔH_s^{Mix} is the partial molar mixing enthalpy of solvent, δ_s (δ_p) the solubility parameter of solvent (polymer), and V_s is the partial molar volume of solvent.

Combination of Eqs. (6.3) and (6.4) yields the well-known Flory–Huggins equation

$$\Delta F_s^{\text{Mix}} = RT[\ln \Phi_s + (1-1/m)\Phi_p + \chi\Phi_p^2] \tag{6.5}$$

where χ is the Huggins's parameter given by

$$\chi = \frac{1}{Z'}(1 - 1/m) + \frac{V_s}{RT}(\delta_s - \delta_p)^2 \qquad (6.6)$$

For diluted polymer solutions in nonpolar solvents the Flory–Huggins equation generally shows reasonable qualitative agreement with experiments. For a quantitative fit a temperature-independent term χ_0 is introduced instead of the first term of the Huggins parameter χ, hence

$$\chi = \chi_0 + \frac{V_s}{RT}(\delta_s - \delta_p)^2 \qquad (6.7)$$

This fitting parameter appears to be much larger than the original term in order to achieve agreement with experiments! Furthermore, the fitting parameter differs for different permeants and concentrations, thus reducing considerably the predictive power of the Flory–Huggins equation.

Although various theoretical explanations are given for the observed discrepancies,[58] we believe that they are mainly due to the approximations applied.

The Flory–Huggins equation only holds for free polymer chains in *diluted solutions*. In the case of solid polymers, however, we are dealing with polymer network structures or semicrystalline polymers in which solvent take-up is restricted by the resulting strain in the amorphous polymer chains, the ends of which either join in crosslinkages or in crystallites.

Flory and Rehner[59] recognized this problem and introduced an entropy term for elastic deformation of the polymer chains in a rubber-like network structure. They applied earlier work of Kuhn[60] (see Appendix V.2), who derived a distribution function for the distance between the ends of a polymer molecule (displacement length). They assumed that the chain displacement length is dilated during swelling by a factor of $\Phi_p^{-1/3}$, being the isotropic linear extension of the polymer phase. Flory[61] furthermore accounted for the degree of interlinking of the polymer chains and finally obtained the following equation:

$$\Delta F_s^{\mathrm{Mix}} = RT\left[\ln \Phi_s + \left(1 - \frac{2V_s}{fL_{\mathrm{av}}}\right)\Phi_p + \frac{V_s \Phi_p^{1/3}}{L_{\mathrm{av}}} + \chi \Phi_p^2\right] \qquad (6.8)$$

where L_{av} is the average molar volume of polymer chains between two crosslinkages while f is the number of polymer chain ends joining in a common crosslinkage.

Kuhn, and thus Flory, assumed that the coiling of the chains is not affected by neighboring polymer molecules. This assumption holds for dilute polymer solutions, but is certainly incorrect for semicrystalline polymers of high crystallinity; the applied correction is then too small. Because Flory and Huggins assumed that the size of a site in the quasi-lattice is equal to the volume of a solvent molecule, the coordination number Z of the lattice will depend on solvent size and concentration. This effect has not been taken into account by them. As a result, their equation predicts too low an effect of solvent size and in fact is limited by its applicability to single-solvent systems only, which is inherent to the adopted definition of a lattice site (namely, that the volume of a site equals the volume of a solvent molecule).

In the derivation of the Scatchard–Hildebrand (solubility parameter) equation, applied by the authors for predicting enthalpy effects, the mutual binary interaction of two compounds (a_{ij}) is related to the respective pure-component interactions (a_{ii} and a_{jj}) by the arbitrary Berthelot equation[58]:

$$a_{ij} = (a_{ii} \cdot a_{jj})^{1/2} \tag{6.9}$$

As a result, the binary interaction parameters are eliminated (see Section 6.3), suggesting that the enthalpy of mixing can be predicted from pure-compound properties, which in general is not true. Due to the assumption of Eq. (6.9) the solubility parameter equation can predict only positive enthalpies, while interactions stronger than dispersion forces and weak polar forces may result in negative enthalpies. It is surprising that, in spite of these fundamental objections, the solubility parameter model is still widely used for predicting enthalpies.

6.2. MODIFIED EQUATION FOR THE PARTIAL MOLAR ENTROPY OF MIXING

By applying Huggins's procedure[50] we derived (see Appendix V) a new equation for the partial molar entropy of mixing of a swollen polymer system. We considered a system of polymer molecules that are partly crystallized and partly in the amorphous state and assumed that no crystals dissolve during swelling. It has been shown experimentally that the permeating molecules are only present in the amorphous polymer phase[62–64] while the crystallites form a separate phase. Therefore, only the first phase need be considered. The problem is then to count the number of possible configurations of the segments of the amorphous polymer chains, which all start on the surface of certain crystallites and end on the surface of certain crystallites without defining the position of these crystallites. This model transforms into that of a crosslinked amorphous polymer if the volume of the crystallites vanishes.

The quasi-lattice is based on a variable coordination number Z_{av}, which equals the volume average of the individual coordination numbers Z_i of the polymer chains and permeants concerned. The permeants are assumed to consist also of segments having the same size as those of the polymer chains, i.e., the size of a site, which can be chosen arbitrarily. This procedure enables the derivation of an equation for multicomponent mixtures.

Unlike Flory and Huggins, we attributed a restricted number of possible positions to the segments in a molecule; this number is called the flexibility and depends on the molecular structure (see Fig. 37). The average flexibility of the polymer segments (y) is reduced by swelling. We assumed in our model that then part of the segments in a polymer chain lose their flexibility completely (hence $y=1$), while the other segments retain their original flexibility.

Flory and Rehner[59] argued that the average distance between the chain ends (s_0) increases due to swelling by a factor $\Phi_a^{-1/3}$, being the isotropic linear extension of the amorphous polymer phase, where Φ_a is the volume fraction of the amorphous polymer. According to Kuhn[60] the original value of s equals the square root of the average amorphous polymer chain length L_{av}. For highly crystalline polymers, however, it is expected that s is proportional to L_{av}. As shown in Appendix V.2, both relationships result in almost the same entropy contribution for elastic strain due to swelling.

The model described yields a partial molar entropy contribution of mixing:

$$\frac{\Delta S_i^{\mathrm{Mix}}}{R} = -\ln\left(\frac{Z_i}{Z_{av}}\Phi_i\right) - 1 + \frac{V_i}{V_{av}}\frac{Z_i}{Z_{av}}\Phi_s - \frac{V_i}{C_L}\Phi_a^{2/3} \tag{6.10}$$

where

$$Z_{av} = \sum \Phi_i Z_i, \qquad \sum \Phi_i = 1,$$

$$\Phi_s/V_{av} = \sum \Phi_i/V_i, \qquad \sum \Phi_i = \Phi_s$$

$$C_L = \text{elastic strain factor} \approx 3\left(\frac{L_{av}-s_0}{s_0}\right)\left(\frac{y}{y-1}\right)v$$

$\Delta S_i^{\mathrm{Mix}}$ is the partial molar entropy of mixing of compound i, Φ_i (Φ_s) the volume fraction of compound i (all permeants) in the amorphous polymer phase, Φ_a the volume fraction of polymer in the amorphous polymer phase, V_i the partial molar volume of compound i, Z_i the coordination number of compound i, y the flexibility of amorphous polymer chains before swelling, and v is the molar volume of a molecular segment (site).

6.3. PARTIAL MOLAR ENTHALPY OF MIXING

The inaccuracy in the solubility or swelling equations is caused to a large extent by the partial molar enthalpy of mixing, for which no reliable theoretical model is available. Therefore, we consider it more extensively in this chapter.

In 1906 van Laar[65] gave a treatment of the vapor pressure of binary liquid mixtures based upon the van der Waals equation. The relation between the van der Waals a for the mixture and a_{ii} and a_{jj} for the pure components, respectively, is, according to van der Waals,[66]

$$a = \theta_i^2 a_{ii} + 2\theta_i \theta_j a_{ij} + \theta_j^2 a_{jj}$$

and for the corresponding values of the van der Waals b,

$$b = \theta_i b_i + \theta_j b_j$$

where θ_i (θ_j) is the number of molecules i (j) per mole mixture and a_{ij} the interaction of molecules i and j.

According to Van Laar the heat of mixing Δh^{Mix} for unexpanded liquids ($V = b$) is then given by

$$\Delta h^{\text{Mix}} = \theta_i \frac{a_{ii}}{b_i} + \theta_j \frac{a_{jj}}{b_j} - \frac{a}{b}$$

Combination of these equations yields

$$\Delta h^{\text{Mix}} = \frac{\theta_i b_i \theta_j b_j}{\theta_i b_i + \theta_j b_j} \left(\frac{a_{ii}}{b_i^2} + \frac{a_{jj}}{b_j^2} - 2\frac{a_{ij}}{b_i b_j} \right)$$

which, after partial differentiation with respect to θ_i, results in the partial molar heat of mixing of compound i (ΔH_i^{Mix}):

$$\Delta H_i^{\text{Mix}} = \frac{b_i \theta_j^2 b_j^2}{(\theta_i b_i + \theta_j b_j)^2} \cdot \left(\frac{a_{ii}}{b_i^2} + \frac{a_{jj}}{b_j^2} - 2\frac{a_{ij}}{b_i b_j} \right)$$

Lorentz and van Laar[67] in 1925 proposed replacing b_i by the partial molar volume V_i and setting

$$\frac{\theta_j b_j}{\theta_i b_i + \theta_j b_j} = \Phi_j = \text{volume fraction of compound } j$$

This results in the expression

$$\Delta H_i^{\mathrm{Mix}} = V_i \Phi_j^2 \left(\frac{a_{ii}}{V_i^2} + \frac{a_{jj}}{V_j^2} - 2\frac{a_{ij}}{V_i V_j} \right) \tag{6.11}$$

A simplification suggested by van Laar and widely accepted is the introduction of the previously mentioned Berthelot equation [Eq. (6.9)], which transforms Eq. (6.11) into

$$\Delta H_i^{\mathrm{Mix}} = V_i \Phi_j^2 \left(\frac{a_{ii}^{1/2}}{V_i} - \frac{a_{jj}^{1/2}}{V_j} \right)^2 \tag{6.12}$$

A final fundamental modification, extending original work of Hildebrand, was derived by Scatchard in 1931.[58] He derived the partial molar energy of mixing $\Delta E_i^{\mathrm{Mix}}$ in the form

$$\Delta E_i^{\mathrm{Mix}} = V_i \Phi_j^2 (\delta_i - \delta_j)^2 = \Delta H_i^{\mathrm{Mix}} - P\,dV \approx \Delta H_i^{\mathrm{Mix}}$$

representing the Scatchard–Hildebrand equation [Eq. (6.4)], ignoring contraction effects. In this equation δ_i (δ_j) is the solubility parameter of compound i (j) given by

$$\delta_i = \frac{a_i^{1/2}}{V_i} = \left(\frac{E_i^{\mathrm{vap}}}{V_i} \right)^{1/2}$$

while E^{vap} is the energy per mol for isothermal vaporization of pure liquid to the ideal gas.

The above equation was extended for multicomponent mixtures to read

$$\Delta H_i^{\mathrm{Mix}} = V_i (\delta_i - \delta_{\mathrm{av}})^2 \tag{6.13}$$

where

$$\delta_{\mathrm{av}} = \sum_{j=1}^{n} \Phi_j \delta_j, \qquad \sum_{j=1}^{n} \Phi_j = 1$$

Although the mixing enthalpy calculated by means of these equations shows considerable deviations from experimental observations, they are widely applied. Several authors attribute the deviations to the existence of different types of interactions, such as contributions for dispersion interactions, for polar forces, or for hydrogen bonds, which incited Hansen and Beerbower[68] among others to propose the following modification:

$$\Delta H_i^{\mathrm{Mix}} = V_i \Phi_j^2 [(\delta_i^{\mathrm{d}} - \delta_j^{\mathrm{d}})^2 + (\delta_i^{\mathrm{p}} - \delta_j^{\mathrm{p}})^2 + (\delta_i^{\mathrm{h}} - \delta_j^{\mathrm{h}})^2] \qquad (6.14)$$

where δ^{d} is the contribution for dispersion interactions, δ^{p} for polar forces, and δ^{h} for hydrogen bonds.

In general no real improvement is observed with this equation, however (see Section 7.3). In our opinion the main reason for the deviations is the elimination of the binary interaction parameters (a_{ij}) by the Berthelot equation (6.9) in the original equations of van Laar and of Scatchard and Hildebrand.

Mixing rules applied, e.g., in predictive computing programs for activity coefficients by means of the group contribution theory (ASOG, etc.)[56] are in principle also based on the Berthelot equation. In order to compensate for the deviations, another binary interaction parameter (l_{ij}) has been introduced in these programs, where

$$a_{ij} = (1 - l_{ij})(a_{ii} a_{jj})^{1/2} \qquad (6.15)$$

It proves to be impossible, however, to correlate the parameter l_{ij} for non-homologous compounds, thus considerably reducing the predictive power. In spite of the many successful applications of these computer programs, it must be emphasized that there is still a lack of fundamental theoretical knowledge on liquid–liquid and polymer–fluid interactions. The well-known correlation techniques fail in cases where specific interactions (hence heat effects) are involved, so fundamental research is required in order to achieve a reliable model for predicting the heat of mixing. Meanwhile we recommend application of Eq. (6.11) for calculating the partial molar enthalpy of mixing. In Appendix VI this equation has been extended for multicomponent mixtures to read

$$\boxed{\Delta H_i^{\mathrm{Mix}} = V_i \left[\sum_{\substack{j=1 \\ j \neq i}}^{n} \Phi_j (1 - \Phi_i)\, \delta_{ji}^2 - \sum_{\substack{j=1 \\ j \neq k \neq i}}^{n} \Phi_j \Phi_k\, \delta_{jk}^2 \right]} \qquad (6.16)$$

where δ_{jk}^2 is a binary mixture parameter given by

$$\delta_{jk}^2 = \frac{a_{jj}}{V_j^2} + \frac{a_{kk}}{V_k^2} - 2\,\frac{a_{jk}}{V_j V_k} \qquad (6.17)$$

For polymers the latter parameter can be determined experimentally from a swelling experiment.

6.4. THE MODIFIED FLORY–HUGGINS EQUATION

It was pointed out in Section 6.1 that the partial molar entropy equations of Flory and Huggins do not account for preferential orientations, nor did we in the derivation of Eq. (6.10) in Section 6.2. Therefore, these equations are in fact only valid for athermal systems.

For the time being we ignore the effect of preferential orientation on entropy for cases in which a heat effect does take place and calculate the partial molar Gibbs free energy of mixing by just adding the partial molar (athermal) entropy of mixing [Eq. (6.10)] and the partial molar enthalpy of mixing. This approach yields

$$\boxed{\frac{\Delta F_i^{\mathrm{Mix}}}{RT} = \ln\left(\frac{Z_i}{Z_{\mathrm{av}}}\Phi_i\right) + 1 - \frac{V_i}{V_{\mathrm{av}}}\cdot\frac{Z_i}{Z_{\mathrm{av}}}\Phi_{\mathrm{s}} + \frac{V_i}{C_{\mathrm{L}}}\Phi_{\mathrm{a}}^{2/3} + \frac{\Delta H_i^{\mathrm{Mix}}}{RT}} \qquad (6.18)$$

In this equation $\Delta H_i^{\mathrm{Mix}}$ can be given by Eq. (6.16).

Apart from the elastic strain term, Eq. (6.18) transforms for a single solvent and equal coordination numbers of the solvent and polymer into the Flory–Huggins equation.

6.5. SOLUBILITY OF GASES IN POLYMERS

As mentioned earlier in Section 3.5, a polymer can be either in the rubbery state or in the glassy state, depending on whether it is above or below its glass-transition temperature (T_g), respectively.

The sorption coefficient of gases in "rubbery" polymers is either constant or increases with pressure. Sorption is assumed to take place homogeneously throughout the whole amorphous polymer phase. The increasing sorption coefficient is generally attributed to "plasticization" of the polymer, causing higher flexibility and mobility of the chains, resulting in increased gas solubility.[21]

The specific volume of a rubbery polymer decreases gradually with decreasing temperature. Below the glass-transition temperature the gradual volume reduction becomes suddenly smaller; the polymer then transforms into the glassy state. This effect is depicted in Fig. 3. It is assumed that in this (amorphous) state the flexibility and mobility of the polymer chains reduce to almost zero, so that they become rigid and cannot follow the volume reduction imposed by cooling. As a result "free volume" is created between the chains, explaining the reduced thermal expansion coefficient. According to the dual-mode sorption theory, this free volume represents the microvoids into which gases may adsorb according to a Langmuir type of isotherm.[36,37]

The sorption coefficient of gases in glassy polymers decreases with increasing pressure. Furthermore, the solubility of a gas is reduced by the presence of other gases. Both effects are shown clearly in Fig. 8, which reflects experimental data of Sanders and Koros.[39] These effects are explained by the limited adsorption capacity of the microvoids as predicted by the Langmuir contribution in the dual-mode sorption equation (3.17). The "free volume" as well as the value of T_g increase with cooling rate of the polymer and decrease with time. In fact they represent nonequilibrium

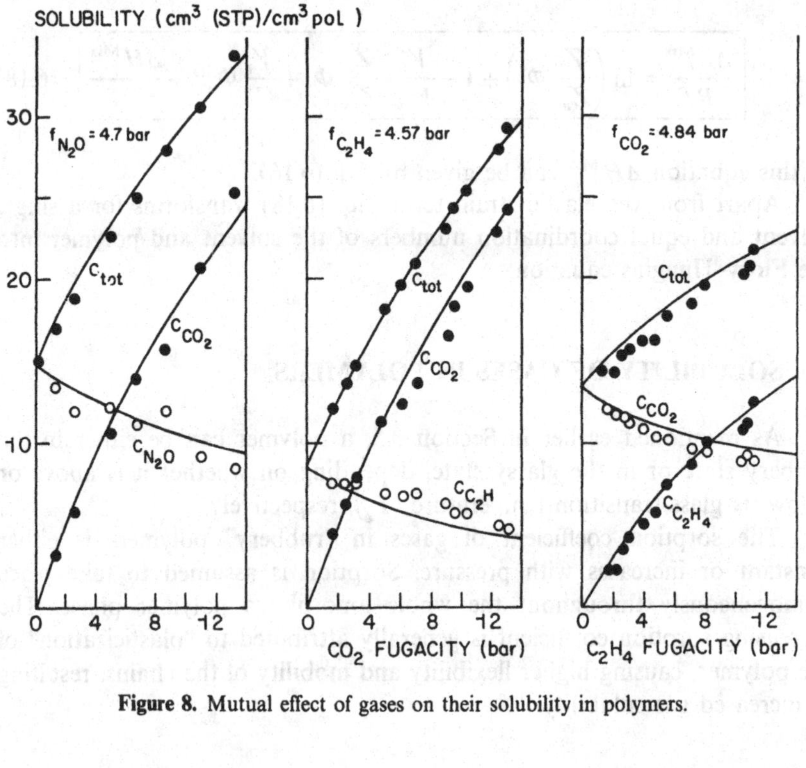

Figure 8. Mutual effect of gases on their solubility in polymers.

properties of the polymer. This is expressed by an exothermal heat effect around T_g in differential scanning calorimetric measurements during heating.[21]

In addition, the hysteresis effects usually observed with glassy polymers in adsorption–desorption experiments with, e.g., CO_2[21,69] point to nonequilibrium conditions.

It was argued above in Section 4.2 that the existence of microvoids has never been shown and is very unlikely from the viewpoint of thermodynamics. In our opinion the deviating behavior of different polymers is due to differences in their solid state structure. Let us consider the structure of the various types of polymers and their properties.

Thermoplasts are polymers which melt upon heating and crystallize (at least partly) upon cooling down. The polymer molecules are partly incorporated in the crystallites and are partly amorphous, thus connecting the individual crystallites and forming a matrix of alternating crystalline and amorphous polymer (semicrystalline polymer; see Fig. 36). Polyolefins which are in the "rubbery" state at room temperature exhibit high crystallinity with large crystallites. Below the melting point the crystallinity hardly depends on temperature at all. This must be attributed to the high heat of crystallization of the polyolefins (200–300 kJ/kg)[70] and the great flexibility and mobility of the amorphous polymer chains. The crystallinity is restricted mainly by chain crossings and branching.

For polyvinyl chloride, a "glassy" polymer at room temperature, a heat of crystallization of between 30 and 130 kJ/kg is quoted.[70] The reported crystallinities, which are determined indirectly from X-ray diffraction, density measurements, and calorimetric measurements, scatter considerably and depend to a large extent on temperature. Furthermore, the crystallites are so small that their existence could be shown only indirectly via small-angle neutron scattering techniques, creating reflection by absorbing a suitable reflecting compound in the amorphous polymer phase.[71]

Generally, long-chain polymer molecules tend to align along each other; this process is accompanied by an exothermic heat effect.

In our opinion local orientation of polymer chains takes place also in glassy polymers, but this effect is counteracted, however, by the thermal motion of the chains. It is expected that the heat dissipation involved in these orientations is relatively small, so that their stability is low and hence depends strongly on temperature. Therefore, the formation of these local orientations extends over a large temperature range, while their equilibrium concentration increases with decreasing temperature.

Below the glass-transition temperature the limited flexibility and mobility of the polymer chains prevent the polymer from reaching the equilibrium concentration of local orientations. As a consequence, the polymer density (which is larger for the aligned polymer chains) and the glass-transition temperature are not constants but depend on the cooling rate of the polymer, the temperature, and the time, as observed experimentally. This theory furthermore explains the reduced thermal volumetric expansion coefficient below T_g. Above this temperature the volumetric reduction during cooling down is partly due to the proceeding alignment of the polymer chains and partly due to the thermal volumetric reduction. Below T_g the alignment is retarded or even stopped, so the overall thermal volumetric expansion coefficient is smaller, as observed experimentally.

In general, sorption takes place in the nonaligned (amorphous) polymer phase.[62–64] Therefore, sorption decreases with increasing concentration of local orientations. Furthermore, such orientations act as crosslinks for the nonaligned polymer chain parts, which may restrict solubility, as will be shown in Section 6.6. However, owing to the limited stability of the local orientations some of them may dissolve in the presence of highly soluble permeants, causing the hysteresis phenomena often observed experimentally.[21,69]

6.6. THE MODIFIED FLORY–HUGGINS EQUATION FOR GASES

As pointed out in Chapter 1 gas (or vapor) molecules dissolved in a polymer are in such close contact with each other and with the polymer molecules that they must be considered as in the liquid phase. Consequently, the sorption of gases in polymers can also be described by the modified Flory–Huggins equation (6.18), which for isothermal and isobaric systems can be written as

$$\ln(a_i) = \ln\left(\frac{Z_i}{Z_{av}}\Phi_i\right) + 1 - \frac{V_i}{V_{av}}\frac{Z_i}{Z_{av}}\Phi_s + \frac{V_i}{C_L}\Phi_a^{2/3} + \frac{\Delta H_i^{Mix}}{RT} \qquad (6.19)$$

This equation contains a large number of physical and thermodynamic parameters, the values of which are unknown in general. However, because the gas solubility in polymers is relatively low, so that Φ_s $(\Phi_i) \ll 1$, Eq. (6.19) can be transformed into a less complicated solubility equation as shown below.

The activity of a gas (vapor) can be expressed by[56]

$$a_i = f_i/f_{sat,i} = \zeta_i P_i/\zeta_{sat,i} P_{sat,i} \qquad (6.20)$$

where ζ is the fugacity coefficient and the subscript sat denotes saturation.

The solubility of a gas is expressed by Henry's law:

$$C_i = {}^{\infty}H_i P_i \tag{6.21}$$

where ${}^{\infty}H_i$ is the Henry coefficient in $\text{cm}^3/(\text{cm}^3 \cdot \text{bar})$. Theoretically, Eq. (6.21) holds only for infinity low gas pressures,[56] hence

$$^{\infty}H_i = \lim_{P_i \to 0} (C_i/P_i) \tag{6.22}$$

Remembering that for $x \ll 1$ we may write

$$(1+x)^n = 1 + nx$$

we derive in Appendix VII from Eqs. (6.19)–(6.22) that

$$C_i = \frac{^{\infty}H_i f_i}{(1-\Phi_s)} \exp(B_i \Phi_s) \tag{6.23}$$

The relationship between C_i and Φ_i is given by

$$\Phi_i = \frac{V_i}{\phi_a} C_i \bigg/ \left(c + \frac{V_i}{\phi_a} C_i \right) \tag{6.24}$$

where c is the factor for converting normal gas volume to moles [22,400 cm^3 (STP)/mol].

The parameter B_i expresses nonideal behavior of the polymer–gas mixture and is given by

$$B_i = \left\{ \frac{V_i}{9v} \cdot \frac{[2f(s)-1]}{f(s)^2} + 2\frac{V_i}{RT}\delta_{ip}^2 + \frac{V_i}{V_{av}} \cdot \frac{Z_i}{Z_{av}} \right\} \tag{6.25}$$

where

$$f(s) = \frac{(L_{av} - s_0)}{s_0} \cdot \frac{y}{(y-1)} \tag{6.26}$$

being a pure polymer property.

The first term between the curly brackets in Eq. (6.25) represents a contribution for elastic strain of the amorphous polymer chains, the second term that for the heat of mixing, and the third term for the size and shape of the compounds involved.

If there is a large difference between L_{av} and s_0, as is the case with rubbers and polyolefins (see Appendix VIII, elastic strain factor), the contribution for elastic strain of the polymer chains on permeant activity is only of minor importance. If, however, s_0 approaches L_{av}, as may be the case, for example, with highly crosslinked (resin-type) structures, the contribution for elastic strain may become overriding.

The numerical value of the last two terms between the curly brackets of Eq. (6.25) is usually less than 2 for gases in polymers. Consequently, B_i is positive if the ratio s_0/L_{av} is smaller than about 0.8 and negative for larger ratios (see Fig. 9). In the former case Eq. (6.23) predicts an increasing solubility coefficient with pressure (as observed with "rubbery" polymers like elastomers and polyolefins) while in the latter case a

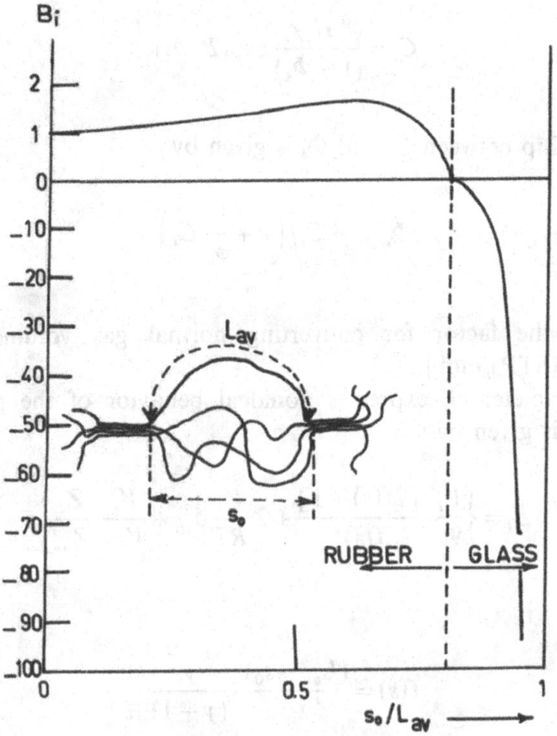

Figure 9. Effect of chain stretching on B_i.

decreasing solubility coefficient with increasing pressure is predicted, as observed with "glassy" polymers. Consequently, the modified Flory–Huggins equation is capable of predicting the gas solubility of rubbery polymers as well as that of glassy polymers. The difference in behavior is explained by different degrees of elastic strain of the amorphous polymer chains due to gas sorption. Hence, there is no need for the arbitrary assumption of the presence of microvoids in glassy polymers, into which gas adsorption takes place according to a Langmuir type of isotherm, as applied in the dual-mode sorption model.

7

COMPARISON AND EXPERIMENTAL CHECK OF THE SOLUBILITY EQUATIONS

A large number of swelling experiments with polyethene and isotactic polypropene have been performed in a great variety of single nonpolar organic solvents such as *n*-paraffines, alkylbenzenes, and cyclanes; these data are used to check and compare the various swelling equations. Furthermore, swelling experiments with natural rubber in various mixtures of toluene and cyclohexanone and in mixtures of carbon tetrachloride and isooctane (2,2,4-trimethylpentane) were performed and published by Paul *et al.*[72] These data are used to check the validity of Eq. (6.18) for swelling of polymers in solvent mixtures.

Finally, as an example of an extremely polar system, swelling experiments were executed with cellulose acetate in ethanol–water mixtures. The required parameters for calculating equilibrium swelling, such as polymer crystallinity, elastic strain factor, coordination numbers, and partial molar heat of mixing, were determined as described in Appendix VIII. For the partial molar volumes of the compounds involved the respective molar volumes were used, thus ignoring the effect of contraction.

For checking the validity of the gas solubility equation (6.23) we apply sorption data for pure N_2O, CO_2, and C_2H_4 and mixtures thereof in polymethyl methacrylate (PMMA), published by Sanders and Koros.[39] We also use experimental results of Sada *et al.*,[44] who determined the sorption and permeability of CO_2 and CH_4 in polysulfone as a function of temperature and pressure.

The results of the various calculations are considered successively in the following sections.

Table 2. Comparison of Measured and Calculated Swelling of Polyethylene in Various Hydrocarbons

Solvent	Molar volume (cm³)	Solubility parameter (cal/cm³)$^{1/2}$	Z_i/Z_a	Measured sorption (g/100 g)	Swelling (volume fraction solvent in amorphous polymer phase): measured	calculated: Eq. (6.18)	calculated: Flory–Huggins[a] A	Flory–Huggins[a] B	Flory–Huggins[a] C
n-Octane	170.1	7.14	1.475	7.1	0.199	0.208	0.231	0.224	0.214
n-Decane	203.0	7.34	1.380	7.8	0.209	0.209[b]	0.209[b]	0.209[b]	0.209[b]
n-Dodecane	236.1	7.48	1.317	8.1	0.210	0.201	0.187	0.192	0.199
n-Hexadecane	303.4	7.65	1.238	6.7	0.177	0.175	0.146	0.158	0.173
n-Octadecane	336.9	7.71	1.211	5.6	0.160	0.162	0.129	0.144	0.163
Cyclohexane	113.7	7.73	1.633	13.8	0.304	0.340	0.385	0.372	0.354
Methylcyclohexane	133.7	7.43	1.543	12.7	0.289	0.292	0.324	0.313	0.298
Benzene	93.4	8.63	2.900	11.2	0.239	0.240	0.434	0.416	0.390
Toluene	110.0	8.54	2.629	11.1	0.240	0.238	0.397	0.385	0.367
Ethylbenzene	127.7	8.40	2.425	10.6	0.231	0.234	0.365	0.357	0.346
n-Propylbenzene	145.0	8.28	2.267	10.0	0.223	0.228	0.335	0.332	0.327
n-Butylbenzene	162.1	8.04	2.140	9.5	0.213	0.218	0.308	0.307	0.307
Isopropylbenzene	145.0	8.15	2.478	9.7	0.218	0.211	0.336	0.332	0.328
o-Xylene	125.3	8.60	2.188	12.0	0.251	0.251	0.363	0.354	0.341
m-Xylene	127.7	8.45	2.188	11.7	0.250	0.254	0.363	0.356	0.345
p-Xylene	128.3	8.40	2.188	11.8	0.253	0.256	0.364	0.356	0.346
p-Cymene	162.5	8.28	2.140	9.7	0.216	0.219	0.308	0.308	0.308

[a] A, Huggins's parameter = 0, $L_{av} = 297$ cm³; B, Huggins's parameter = 0.25, $L_{av} = 394$ cm³; C, Huggins's parameter = 0.50, $L_{av} = 586$ cm³.

[b] Data used for calculating L_{av} and C_L $\left(\frac{L_{av} - s_o}{s_o} \cdot \frac{y}{y-1} = 7.853\right)$.

7.1. SWELLING OF POLYOLEFINS

The results of swelling experiments with polyolefins in various hydrocarbons are collected in Tables 2 and 3. The volume fractions of solvents indicated are based on the amorphous polymer, assuming that no solvent is present in the crystallites and no crystals dissolve in the solvents.

The elastic strain factor C_L in Eq. (6.18) and the average length of the polymer chains between two crosslinkages L_{av} in Eq. (6.8) were derived from the swelling experiment in n-decane. Owing to the lack of sufficient information, the partial molar enthalpy of mixing, which has only a second-order effect in these systems, was estimated from the solubility parameters using the equation of Scatchard and Hildebrand.

It is noted in Appendix VIII that the ratios of the coordination numbers of n-hydrocarbons in polyethylene can be represented by the ratios of their specific volumes. However, a better fit is achieved by applying a kind of group contribution model, according to which each hydrocarbon group that fits the lattice structure of a polyolefin is counted once, while each group that does not fit it is counted 2.9 times. In Appendix VIII a more extensive treatment of this group contribution model is given.

The various physical and thermodynamic data thus obtained enabled the equilibrium sorption (swelling) of polyethylene in various hydro-

Table 3. Comparison of Measured and Calculated Swelling of Polypropylene in Various Hydrocarbons

Solvent	Molar volume (cm^3)	Solubility parameter $(cal/cm^3)^{1/2}$	Z_i/Z_a	Solvent take-up (g/100 g)	Swelling (volume fraction solvent in amorphous polymer phase) measured	calculated by Eq. (6.18)
Isooctane	173	6.48	1.237	25.9	0.250	0.242
Methylcyclopentane	117	7.33	1.316	57.7	0.426	0.444
Cyclohexane	113.7	7.73[a]	1.129	81.4	0.512	0.512[b]
Methylcyclohexane	133.7	7.43	1.271	57.9	0.427	0.428
Toluene	110	8.54	1.813	39.2	0.335	0.322
Ethylbenzene	127.7	8.40	1.949	30.0	0.279	0.285
Isopropylbenzene	145.0	8.15	2.054	26.1	0.251	0.268
$(o+m+p)$-Xylene	127.1	8.48	1.949	31.6	0.289	0.277

[a] Heat of mixing assumed to be zero.

[b] Data used for calculating C_L $\left(\frac{L_{av}-s_o}{s_o}\cdot\frac{y}{y-1}=9.335\right)$

Table 4. Prediction of the Swelling of Polypropylene with the Flory–Huggins Equation

	Swelling (volume fraction)		
	measured	calculated	
Average chain length L_{av}	—	593	603
Huggins's parameter	—	0.25	0.50
Isooctane	0.250	0.250[a]	0.185
Methylcyclopentane	0.426	0.453	0.358
Cyclohexane	0.512	0.470	0.374
Methylcyclohexane	0.427	0.430	0.340
Toluene	0.335	0.434	0.335[a]
Ethylbenzene	0.279	0.414	0.321
Isopropylbenzene	0.251	0.406	0.318
$(o+m+p)$-Xylene	0.289	0.407	0.315

[a] Data used for calculating L_{av}.

carbons to be calculated with Eq. (6.18). Calculations prove to be in excellent agreement with experiment (see Table 2).

For comparison, the swelling was also calculated by means of the Flory–Huggins–Rehner equation, varying Huggins's parameter from 0 to 0.5. Table 2 shows that it is not possible to fit all data with one single value of this parameter.

Table 3 presents equilibrium swelling results of polypropylene in various hydrocarbons. The swelling experiment in cyclohexane was used to calculate C_L. The coordination numbers were estimated by using the aforementioned group contribution model.

As shown in Table 3, the calculated swelling from the various data are in good agreement with experimental results. The results in Table 4 show that it is impossible to match these swelling experiments with a single value of Huggins's parameter in the conventional Flory–Huggins equation.

7.2. SWELLING OF NATURAL RUBBER IN BINARY SOLVENT MIXTURES

Table 5 shows swelling experiments of natural rubber in mixtures of toluene and cyclohexane and of iso-octane and carbontetrachloride, reported by Paul *et al.*[72]

In Appendix VIII the determination of the required physical and thermodynamic data is given for calculating the equilibrium swelling of the

Table 5. Comparison of Measured and Calculated Swelling of Natural Rubber

In Mixtures of Toluene and Cyclohexanone

Composition solvent phase				Composition swollen polymer phase (volume fraction)			
toluene		cyclohexanone				total solvent content	
volume fraction	activity	volume fraction	activity	toluene calculated	cyclohexanone calculated	calculated	measured
0	0	1.00	1.00	0	0.7190	0.719[a]	0.719
0.20	0.2035	0.80	0.8086	0.1562	0.5760	0.732	0.730
0.50	0.4997	0.50	0.5194	0.3940	0.3709	0.765	0.763
0.80	0.7961	0.20	0.2164	0.6423	0.1540	0.796	0.796
1.00	1.0000	0	0	0.8100	0	0.810[a]	0.810

In Mixtures of Isooctane and Carbontetrachloride

Composition solvent phase				Composition swollen polymer phase (volume fraction)			
isooctane		carbontetrachloride				total solvent content	
volume fraction	activity	volume fraction	activity	isooctane calculated	tetra calculated	calculated	measured
1.00	1.0000	0	0	0.6610	0	0.661[a]	0.661
0.75	0.6615	0.25	0.4486	0.4924	0.2341	0.727	0.724
0.50	0.4316	0.50	0.6948	0.3348	0.4390	0.774	0.768
0.25	0.2343	0.75	0.8577	0.1781	0.6283	0.806	0.801
0	0	1.00	1.0000	0	0.8390	0.839[a]	0.839

[a] Data used for calculating Z_i/Z_s. Temperature 30°C.

rubber in the various hydrocarbon mixtures using Eq. (6.18). The respective coordination numbers were derived from the measured swelling in the pure components. The results in Table 5 show excellent agreement between calculated and measured swelling in the binary mixtures, indicating that Eq. (6.18) is capable of predicting equilibrium swelling of polymers in multicomponent solvent mixtures.

7.3. SWELLING OF CELLULOSE DIACETATE IN MIXTURES OF ETHANOL AND WATER

The previous experiments only involve dispersion interactions or weak polar forces, so that the enthalpy contribution, estimated by the solubility parameter model, Eq. (6.4), had only a second-order effect.

Table 6. Binary Mixing Parameters

System	δ_{ij}^2		
	Solubility parameter	Hansen	Experimental
MEK[a]–toluene	0.1369	17.04	2.621
MEK–bright stock	0.4543	26.10	7.892
MEK–elastomer	0.1063	—	0.0
Toluene–bright stock[b]	0.0924	1.484	3.419
Toluene–elastomer	0.4844	—	5.605
Bright stock–elastomer	1.000	—	3.810
EtOH–water	109.2	175.8	15.05
EtOH–cellulose acetate	3.24	17.23	—
Water–cellulose acetate	157.0	272.0	32.26

[a] Methyl-ethyl-ketone.
[b] Bright stock lubricating oil.

In the system ethanol–water–cellulose diacetate, however, strong polar forces and hydrogen bonds are involved. For this system the enthalpy of mixing cannot be described by means of the solubility parameter. This follows clearly from Table 6, in which the mutual binary enthalpies of mixing of these three components, calculated from the corresponding swelling experiments, are compared with those calculated from the Scatchard–Hildebrand equation and from the Hansen equation. We therefore applied Eq. (6.16) to estimate the enthalpy contributions for the three component mixtures, using the swelling or mixing data of the respective binary systems. The reader is referred to Appendix VIII for the derivation of the other data required.

The results in Fig. 10 show excellent confirmation of measured and calculated swelling of the cellulose diacetate in mixtures of water and ethanol. Consequently, it may be concluded that at least a good semi-empirical description of the equilibrium swelling of strong polar multicomponent systems can be achieved with Eq. (6.18). It must be emphasized that quantitative prediction of the behavior of such systems is only possible if a reliable relation is available for predicting the partial molar enthalpy of mixing.

7.4. SOLUBILITY OF N_2O, CO_2, AND C_2H_4 IN PMMA

Sanders and Koros[39] have published experimental sorption data for pure N_2O, CO_2, and C_2H_4 and mixtures thereof in polymethyl

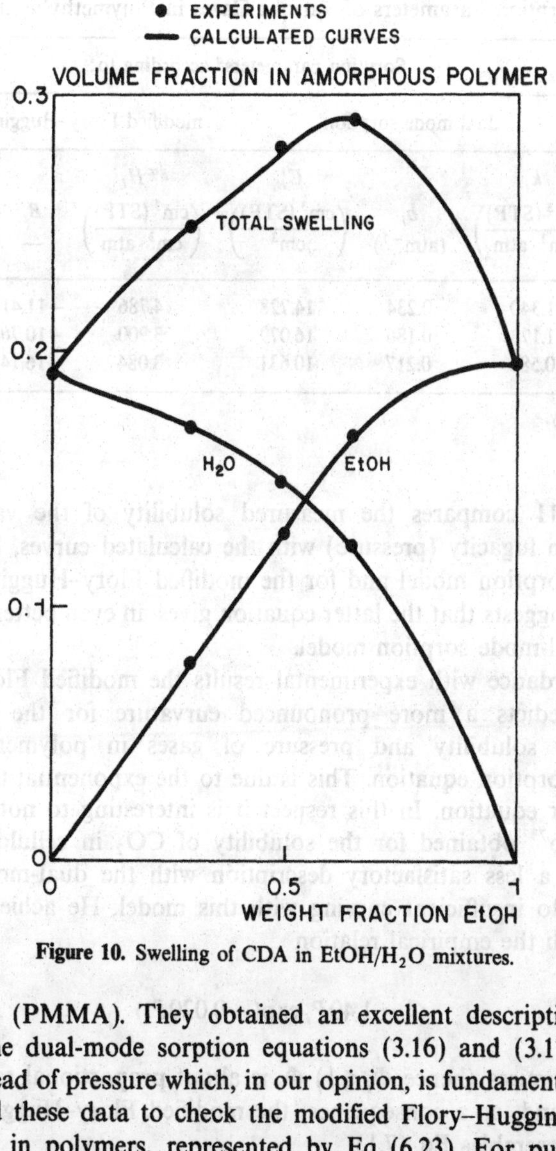

Figure 10. Swelling of CDA in EtOH/H_2O mixtures.

methacrylate (PMMA). They obtained an excellent description of these data with the dual-mode sorption equations (3.16) and (3.17), applying fugacity instead of pressure which, in our opinion, is fundamentally correct.

We used these data to check the modified Flory–Huggins model for gas mixtures in polymers, represented by Eq. (6.23). For pure gases Φ_s transforms into Φ_i in this equation. According to Eq. (6.23) two parameters are required in order to calculate the solubility, namely, ${}^{\infty}H_i$ and B_i. However, to calculate Φ_i the parameter V_i/Φ_a is also required, as follows from Eq. (6.24). These parameters are derived from some experimental results obtained by Sanders and Koros and are collected in Table 7, together with the dual-mode sorption parameters derived by the same authors.

Table 7. Sorption Parameters of Various Gases in Polymethylmethacrylate[39]

	Sorption parameters^a according to:					
	dual-mode sorption			modified Flory–Huggins equation		
Gas	k_i $\left(\frac{cm^3\,(STP)}{cm^3 \cdot atm}\right)$	b_i (atm^{-1})	C'_{H_i} $\left(\frac{cm^3\,(STP)}{cm^3}\right)$	$^{\infty}H_i$ $\left(\frac{cm^3\,(STP)}{cm^3 \cdot atm}\right)$	B_i —	V_i/Φ_a (cm^3/mol)
N_2O	1.340	0.234	14.728	4.786	−11.41	60
CO_2	1.173	0.186	16.079	3.900	−10.76	60
C_2H_4	0.586	0.217	10.631	3.084	−16.14	90

^{a} Based on fugacity.

Figure 11 compares the measured solubility of the various gases depending on fugacity (pressure) with the calculated curves, both for the dual-mode sorption model and for the modified Flory–Huggins equation. This figure suggests that the latter equation gives an even better description than the dual-mode sorption model.

In accordance with experimental results the modified Flory–Huggins equation predicts a more pronounced curvature for the relationship between the solubility and pressure of gases in polymers than the dual-mode sorption equation. This is due to the exponential term with Φ_i in the former equation. In this respect it is interesting to note that Stern and Meringo[73] obtained for the solubility of CO_2 in cellulose diacetate above 45°C a less satisfactory description with the dual-mode sorption model, due to insufficient curving with this model. He achieved a much better fit with the empirical relation

$$C_i = 1.40 P_i \exp(-0.020 P_i) \tag{7.1}$$

At lower pressures (hence $\Phi_i \ll 1$) Φ_i is about proportional to the partial pressure P_i and, as a consequence, the modified Flory–Huggins equation (6.23) then resembles Eq. (7.1).

The results in Table 7 suggest that the parameter B_i in the modified Flory–Huggins equation is proportional to V_i/Φ_a, hence to V_i, since the amorphous polymer fraction Φ_a is the same for the three gases examined. This is indeed predicted by Eq. (6.25), if it is remembered that the sum of the last two terms on the right-hand side of this equation amounts to less than 2 for gases in "glassy" polymers, so that B_i is mainly determined by the first term, being a polymer property.

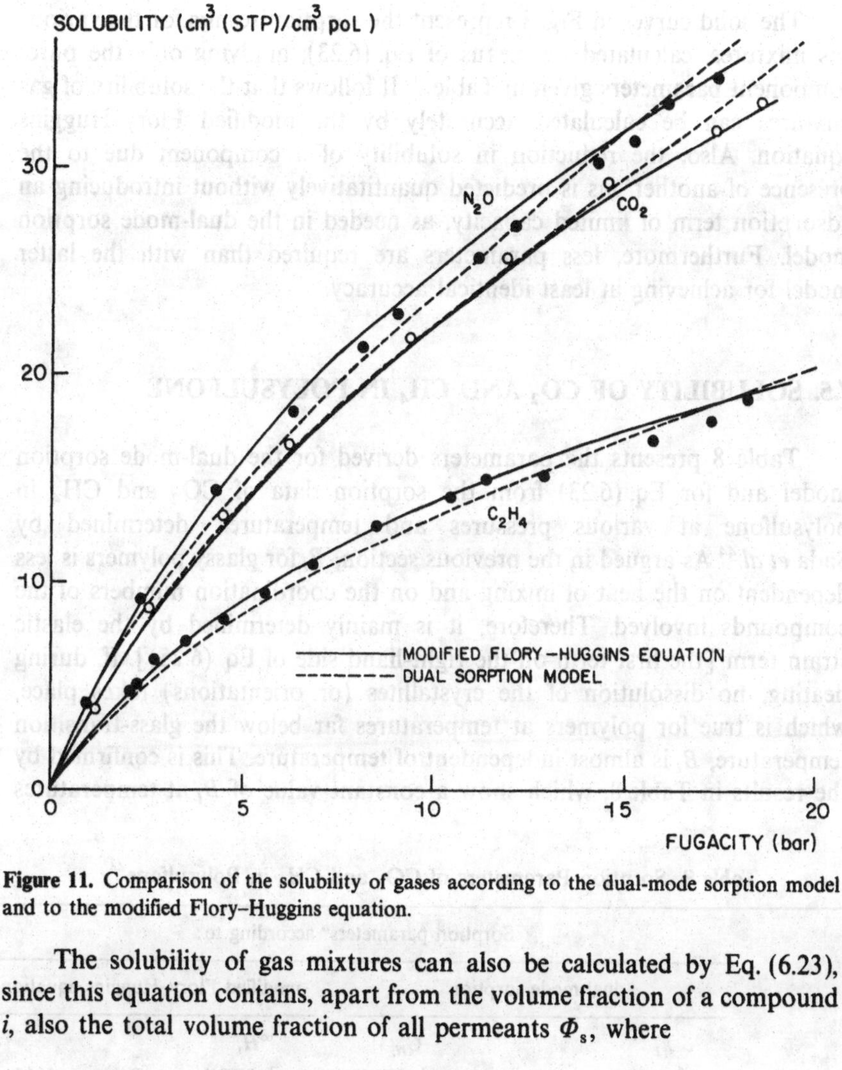

Figure 11. Comparison of the solubility of gases according to the dual-mode sorption model and to the modified Flory–Huggins equation.

The solubility of gas mixtures can also be calculated by Eq. (6.23), since this equation contains, apart from the volume fraction of a compound *i*, also the total volume fraction of all permeants Φ_s, where

$$\Phi_s = \sum_{i=1}^{n} \Phi_i$$

In order to check the validity of this equation for gas mixtures we use experimental data of Sanders and Koros, who measured the solubility of binary gas mixtures in PMMA, depending on the fugacity (pressure) of one of the gases while keeping that of the other one constant. Their results are shown in Fig. 8, which shows that the solubility of the gas, maintained at constant fugacity, decreases with increasing fugacity (hence sorption) of the other gas.

The solid curves in Fig. 8 represent the sorption results for the various gas mixtures, calculated by means of Eq. (6.23), applying only the pure-component parameters given in Table 7. It follows that the solubility of gas mixtures can be calculated accurately by the modified Flory–Huggins equation. Also, the reduction in solubility of a component due to the presence of another gas is predicted quantitatively without introducing an adsorption term of limited capacity, as needed in the dual-mode sorption model. Furthermore, less parameters are required than with the latter model for achieving at least identical accuracy.

7.5. SOLUBILITY OF CO_2 AND CH_4 IN POLYSULFONE

Table 8 presents the parameters derived for the dual-mode sorption model and for Eq. (6.23) from the sorption data of CO_2 and CH_4 in polysulfone at various pressures and temperatures, determined by Sada *et al.*[44] As argued in the previous section, B_i for glassy polymers is less dependent on the heat of mixing and on the coordination numbers of the compounds involved. Therefore, it is mainly determined by the elastic strain term [the first term on the right-hand side of Eq. (6.25)]. If, during heating, no dissolution of the crystallites (or orientations) takes place, which is true for polymers at temperatures far below the glass-transition temperature, B_i is almost independent of temperature. This is confirmed by the results in Table 8, which show a constant value of B_i at temperatures

Table 8. Sorption Parameters of CO_2 and CH_4 in Polysulfone[44]

		Sorption parameters[a] according to:					
		dual-mode sorption			modified Flory–Huggins equation		
Gas	Temp. (°C)	k_i $\left(\frac{cm^3\,(STP)}{cm^3 \cdot atm}\right)$	b_i (atm^{-1})	C'_{H_i} $\left(\frac{cm^3\,(STP)}{cm^3}\right)$	$^{\infty}H_i$ $\left(\frac{cm^3\,(STP)}{cm^3 \cdot atm}\right)$	B_i —	V_i/Φ_a (cm^3/mol)
CO_2	30	0.737	0.199	19.6	4.644	−17.53	60
	35	0.677	0.183	18.5	4.048	−17.53	60
	40	0.630	0.168	17.4	3.544	−17.53	60
	45	0.595	0.155	16.2	3.159	−17.53	60
CH_4	30	0.489	0.121	8.37	1.390	−17.53	60
	40	0.445	0.0976	6.60	1.089	−17.53	60
	50	0.415	0.0785	5.12	0.887	−17.53	60

[a] Based on fugacity.

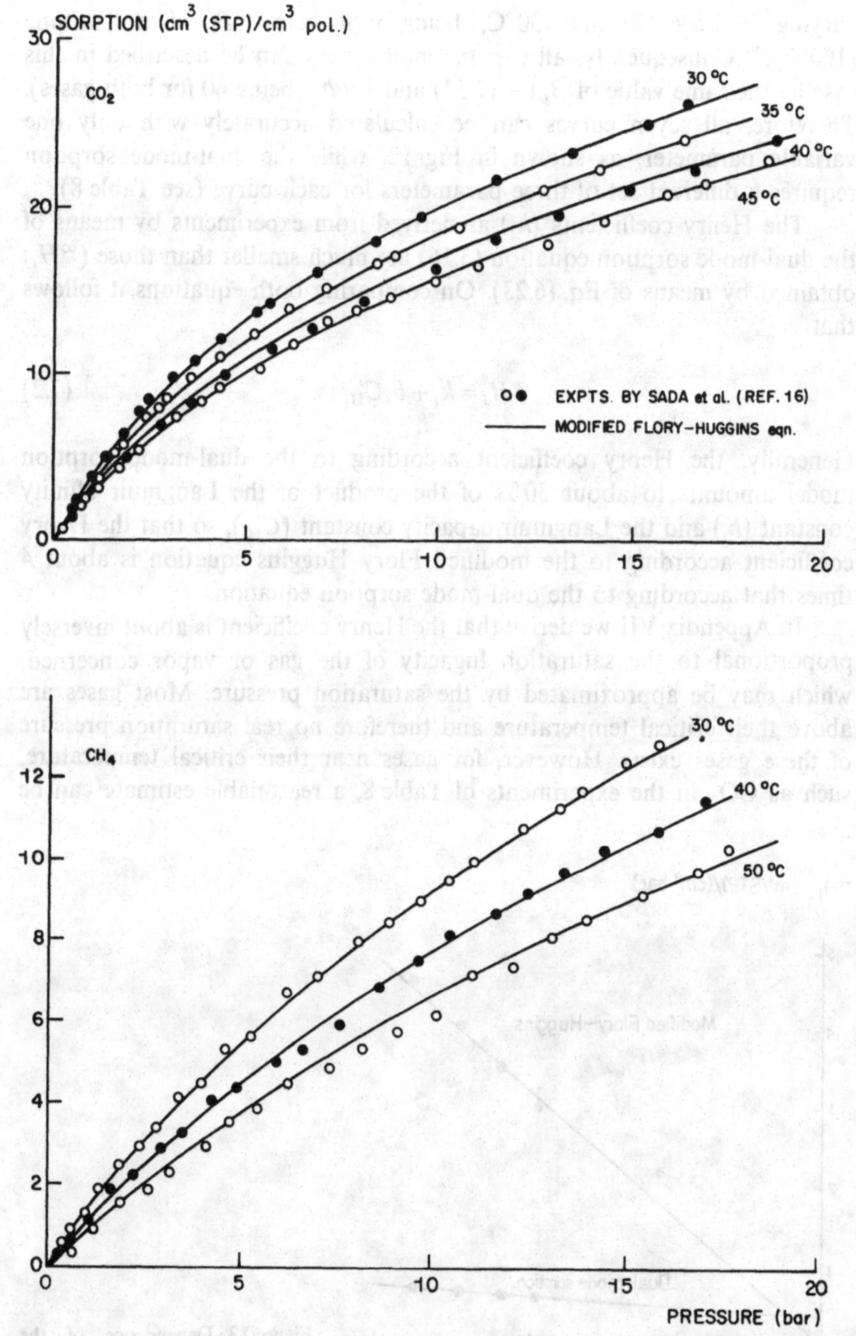

Figure 12. Prediction of gas solubility in polymers by means of the modified Flory–Huggins equation.

varying between 30 and 50°C, being well below T_g of polysulfone (190°C).[44] Consequently, all experimental curves can be described in this case by the same value of B_i (−17.53) and V_i/Φ_a (being 60 for both gases). Therefore, all seven curves can be calculated accurately with only one variable parameter, as shown in Fig. 12, while the dual-mode sorption requires a different set of three parameters for each curve (see Table 8).

The Henry coefficients (k_i) as derived from experiments by means of the dual-mode sorption equation (3.16) are much smaller than those (${}^{\infty}H_i$) obtained by means of Eq. (6.23). On comparing both equations it follows that

$$ {}^{\infty}H_i = k_i + b_i C'_{Hi} \tag{7.2}$$

Generally, the Henry coefficient according to the dual-mode sorption model amounts to about 30% of the product of the Langmuir affinity constant (b_i) and the Langmuir capacity constant (C'_{Hi}), so that the Henry coefficient according to the modified Flory–Huggins equation is about 4 times that according to the dual-mode sorption equation.

In Appendix VII we derive that the Henry coefficient is about inversely proportional to the saturation fugacity of the gas or vapor concerned, which may be approximated by the saturation pressure. Most gases are above their critical temperature and therefore no real saturation pressure of these gases exists. However, for gases near their critical temperature, such as CO_2 in the experiments of Table 8, a reasonable estimate can be

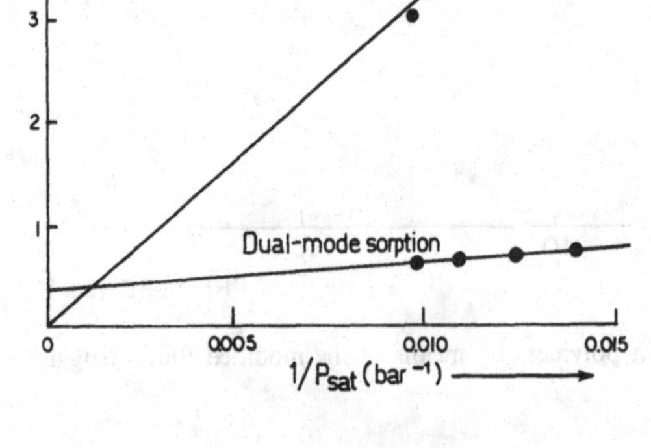

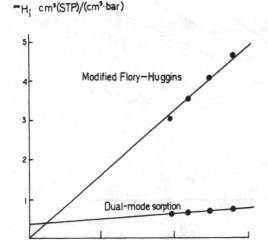

Figure 13. Dependence of the Henry coefficient of CO_2 on saturation pressure.

made by extrapolating the curve $\ln P_{sat}$ versus $1/T$ above that critical temperature.

The Henry coefficients of CO_2 in polysulfone, obtained according to both sorption models, are plotted in Fig. 13 versus the saturation pressure, estimated according to the procedure previously described. It follows that the Henry coefficients according to the modified Flory–Huggins equation (6.23) are indeed inversely proportional to the saturation pressure, while those according to the dual-mode sorption model are not.

8

PREDICTION OF DIFFUSIVITY IN MULTICOMPONENT MIXTURES

The diffusivity of a component in a gas mixture is much less dependent on composition than in a liquid mixture. Therefore, prediction of gas diffusivity is generally easier and more accurate than that of liquids. In this book we consider transport phenomena in polymers and because a gas, dissolved in a polymer, must considered to be in the liquid phase, we restrict ourselves to the diffusivity in liquids and polymers. For more information about gas diffusivity we refer the reader to the literature.[56]

8.1. BASIC DIFFUSIVITY

When predicting diffusivity in multicomponent mixtures we commonly employ arbitrary relationships which relate the diffusivity of a component i in such mixtures to the *basic diffusivities* ${}^{\infty}D_{ij}$ (i.e., the binary diffusivity of i in j at infinite dilution of i) of the components involved. For estimating the basic diffusivities many correlations have been developed, such as the equations of Wilke–Chang, Scheibel, Reddy–Doraiswamy, Lusis–Ratcliff, King–Hsueh–Mao, and many others[56] (see also Table 9). These semi-empirical relations are based on the Stokes–Einstein equation

$$\frac{{}^{\infty}D_{ij}\eta_j}{T} = \text{function of solute size and shape } (R/\sigma_{ij}) \qquad (8.1)$$

in which η_j is the viscosity of the solvent.

For not too high viscosities (e.g., below 50 cP) and if no strong polar interactions occur, reasonably accurate estimates are obtained with all the

Table 9. Diffusivity Correlations[a]

Stokes–Einstein	$^{\infty}\sigma_{ij} = RT/^{\infty}\mathbf{D}_{ij}\eta_j$
Wilke–Chang	$^{\infty}\sigma_{ij} = 1.12 \cdot 10^8 V_i^{0.6}(\Phi M_j)^{-0.5}$ for H_2O $\Phi = 2.6$ for nonassociated solvents $\Phi = 1$
Scheibel	$^{\infty}\sigma_{ij} = 10^8 V_i^{0.33}[1 + (3V_j/V_i)^{0.67}]^{-1}$
Reddy–Doraiswamy	$^{\infty}\sigma_{ij} = 8.3 \cdot 10^7 (V_i V_j)^{0.33} M_j^{-0.5}$ for $V_j \geqslant 1.5V_i$ $^{\infty}\sigma_{ij} = 9.8 \cdot 10^7 (V_i V_j)^{0.33} M_j^{-0.5}$ for $V_j \leqslant 1.5V_i$
Lusis–Ratcliff	$^{\infty}\sigma_{ij} = 10^8 V_j^{0.33}[1.4(V_j/V_i)^{0.33} + V_j/V_i]^{-1}$
King–Hsueh–Mao	$^{\infty}\sigma_{ij} = 1.9 \cdot 10^8 (V_i/V_j)^{0.167}(H_i/H_j)^{0.5}$

[a] T (K), R (J/mol·K), η (cP), $^{\infty}\mathbf{D}_{ij}$ (cm^2/s), V_i (cm^3/mol), M_j (g/mol), H_i (heat of evaporation in J/mol).

correlations mentioned, proving the validity of Eq. (8.1) as shown in Fig. 14.[56] For high viscosities and strong interactions the following modification of this equation is often applied[56]:

$$^{\infty}D_{ij} = \text{constant} \times \eta_j^q \tag{8.2}$$

where q varies between 0.5 and 1, depending on the type of solvent j.

8.2. CONCENTRATED SOLUTE DIFFUSIVITY

Many empirical relationships are used to estimate the diffusivity in binary mixtures from the mutual basic diffusivities ($^{\infty}D_{ij}$ and $^{\infty}D_{ji}$)

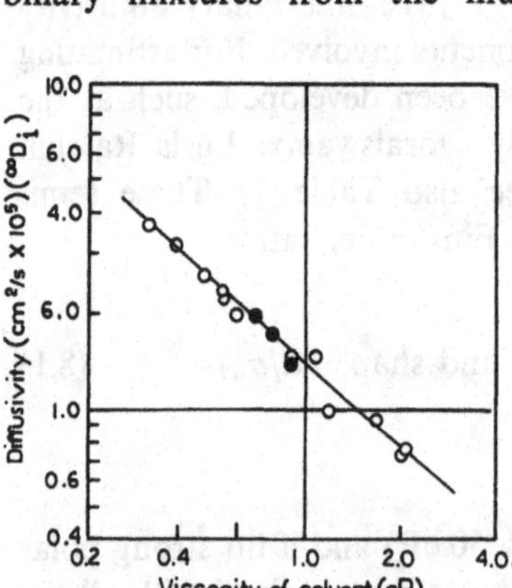

Figure 14. Diffusivity of carbon tetrachloride in various hydrocarbons.

depending on the concentration. It has been reported that a good description is obtained from the Vigne equation for almost ideal mixtures[56]:

$$\ln D_i = x_i \ln {}^{\infty}D_{ji} + x_j \ln {}^{\infty}D_{ij} \tag{8.3}$$

For nonideal mixtures the following modification has been proposed:

$$\ln(D_i \eta_{\mathrm{m}}) = x_i \ln({}^{\infty}D_{ji}\eta_i) + x_j \ln({}^{\infty}D_{ij}\eta_j) \tag{8.4}$$

where m denotes mixture.

The latter equation generally gives only a minor improvement. This can be explained by considering the equation of Andrade[74] which relates mixture viscosity to that of the pure components:

$$\ln \eta_{\mathrm{m}} = x_i \ln \eta_i + x_j \ln \eta_j \tag{8.5}$$

For small differences in molar volume and polarity of the components this equation gives a good description (see also Section 8.4). Combination of the latter two equations yields the original Vigne equation, so that indeed less effect of the viscosity correction in Eq. (8.4) can be expected. With nonideal mixtures a considerable improvement is achieved by application of Eq. (8.4),[75] but then Eq. (8.5) is no longer correct for a proper prediction of viscosity.

According to Eqs. (8.3) and (8.4)

$$D_i = D_j$$

Consequently, these equations predict mutual diffusion coefficients, namely, diffusivities defined by the diffusion equation of Fick, which does not account for convection.

When estimating the diffusivity in a multicomponent mixture Lenoir[76] treats the system as a quasi-binary mixture consisting of component i and all others lumped together as component m, using the equation of Caldwell and Babb[77]:

$$D_{i\mathrm{m}} = x_i {}^{\infty}D_{\mathrm{m}i} + (1 - x_i) {}^{\infty}D_{i\mathrm{m}} \tag{8.6}$$

Lenoir calculates ${}^{\infty}D_{i\mathrm{m}}$ by applying Cullinan's relation[78]:

$$\ln {}^{\infty}D_{i\mathrm{m}} = \sum_{\substack{j=1 \\ j \neq i}}^{n} \frac{x_j}{(1 - x_i)} \ln {}^{\infty}D_{ij} \tag{8.7}$$

For calculating $^{\infty}D_{mi}$ he uses the molar average of the basic diffusivities:

$$^{\infty}D_{mi} = \sum_{\substack{j=1 \\ j \neq i}}^{n} \frac{x_j}{(1-x_i)} \, ^{\infty}D_{ji} \tag{8.8}$$

Lenoir remarks that, although the basic diffusivities $^{\infty}D_{ij}$ are all different, the diffusivities of the components in the mixture trend toward a centralized value. This is indeed predicted by the previous two equations.

Our criticisms concerning the Lenoir equations are their arbitrary character and the fact that *mutual diffusivities* are obtained. The latter is also true of the equation of Vigne, which is caused by the fact that *mutual basic diffusivities* are used to calculate mixture diffusivity. However, according to the diffusion equation of Darken, Prager, and Crank and the modified Maxwell–Stefan equation (see Sections 5.3 and 5.4, respectively) the *coefficient of self-diffusion* in the mixture ($^{*}D_{im}$) rather than the mutual diffusivity (D_{im}) must be employed.

Self-diffusion coefficients, measured by means of radioactive tracers or by the NMR–spin echo method, have been reported for various binary mixtures.[55] Typical examples of such measurements are presented in Fig. 15. It follows that the *self-diffusion coefficient* of a component i in a mixture of components i and j is not limited by the *mutual basic diffusivities* but by the *basic diffusivity* $^{\infty}D_{ij}$ and the *coefficient of self-diffusion of i in pure i*, denoted by $^{*}D_{ii}$ and referred to as the *basic self-diffusivity*.

Obviously, it is more correct to describe the coefficients of self-diffusion by means of the Vigne equation rather than the mutual diffusivities. Consequently Eq. (8.3) transforms into

$$\ln {}^{*}D_{im} = x_i \ln {}^{*}D_{ii} + x_j \ln {}^{\infty}D_{ij} \tag{8.9}$$

Starting from this equation, we derived in Appendix IX the following equation for the self-diffusion in multicomponent mixtures:

$$\ln {}^{*}D_{im} = x_i \ln {}^{*}D_{ii} + \sum_{\substack{j=1 \\ j \neq i}}^{n} x_j \ln {}^{\infty}D_{ij} \tag{8.10}$$

which is fundamentally more correct and less complicated than the Lenoir equations.

Membrane separations usually involve polymers. According to the theory of Flory and Huggins[50,51] a swollen membrane may be regarded as a homogeneous liquid mixture of polymer and permeants and therefore the

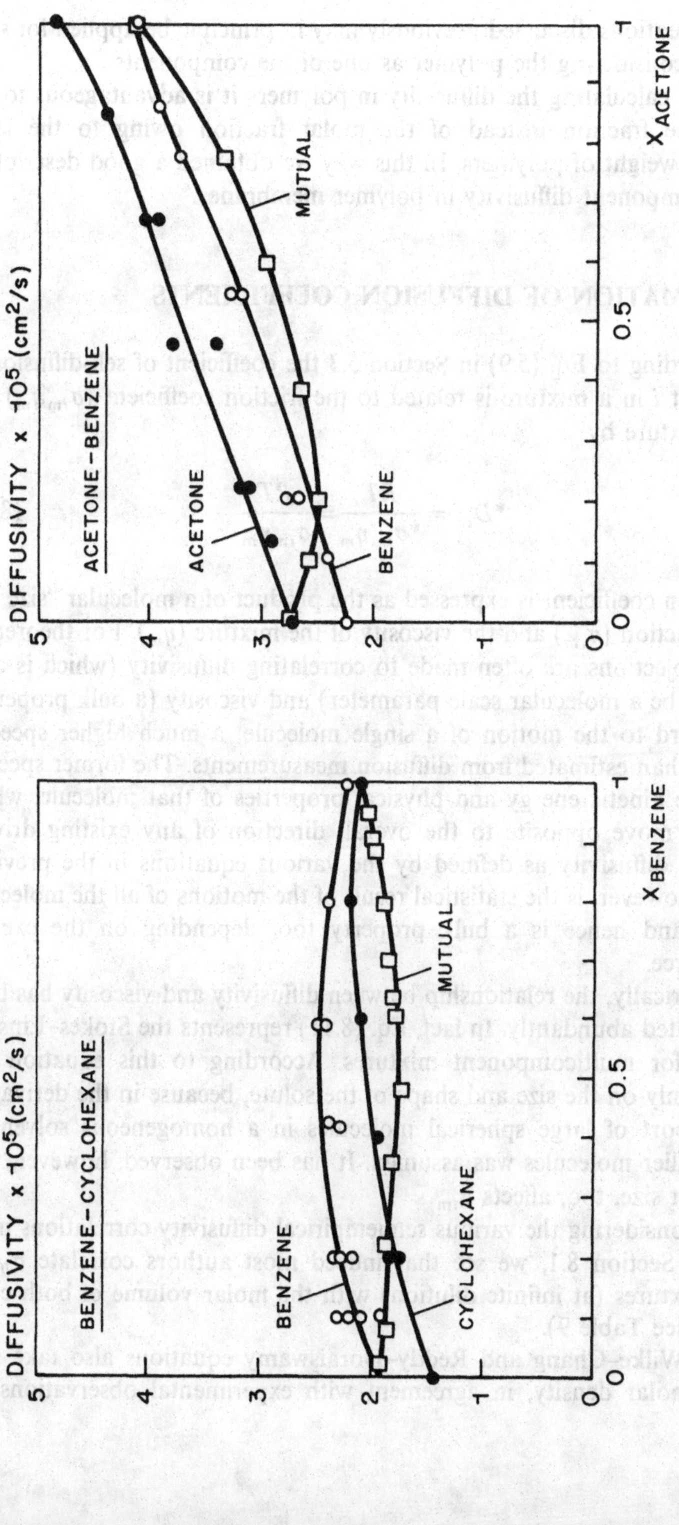

Figure 15. Comparison of self- and mutual diffusivity in binary mixtures.

various equations discussed previously may in principle be applied for such mixtures, considering the polymer as one of the components.

When calculating the diffusivity in polymers it is advantageous to use the volume fraction instead of the molar fraction owing to the large molecular weight of polymers. In this way we obtained a good description of multicomponent diffusivity in polymer membranes.[4]

8.3. ESTIMATION OF DIFFUSION COEFFICIENTS

According to Eq. (5.9) in Section 5.3 the coefficient of self-diffusion of component i in a mixture is related to the friction coefficient ($\sigma_{im}\eta_m$) of i in that mixture by

$$^*D_{im} = \frac{RT}{^*\sigma_{im}\eta_m} = \frac{RT}{\sigma_{im}\eta_m} \tag{8.11}$$

The friction coefficient is expressed as the product of a molecular "size and shape" function (σ_{im}) and the viscosity of the mixture (η_m). For theoretical reasons objections are often made to correlating diffusivity (which is considered to be a molecular scale parameter) and viscosity (a bulk property). With regard to the motion of a single molecule, a much higher speed is observed than estimated from diffusion measurements. The former speed is due to the kinetic energy and physical properties of that molecule, which may even move opposite to the overall direction of any existing driving force. The diffusivity as defined by the various equations in the previous sections, however, is the statistical result of the motions of all the molecules involved and hence is a bulk property too, depending on the exerted driving force.

Empirically, the relationship between diffusivity and viscosity has been demonstrated abundantly. In fact, Eq. (8.11) represents the Stokes–Einstein equation for multicomponent mixtures. According to this equation σ_{im} depends only on the size and shape of the solute, because in the derivation the transport of large spherical molecules in a homogeneous solvent of much smaller molecules was assumed. It has been observed, however, that the solvent size, too, affects σ_{im}.

On considering the various semiempirical diffusivity correlations mentioned in Section 8.1, we see that indeed most authors correlate σ_{ij} for binary mixtures (at infinite dilution) with the molar volume of both components (see Table 9).

The Wilke–Chang and Reddy–Doraiswamy equations also take into account molar density, in agreement with experimental observations. In general σ_{ij} is found to decrease with increasing density of j and increasing molar volume ratio V_j/V_i. Furthermore, the viscosity increases at equal molar volume with increasing density, or at equal density with increasing molar volume. As the diffusivity is inversely proportional to the product of viscosity and σ_{ij}, diffusivity increases less than inversely proportional to solvent viscosity, in agreement with the empirical Eq. (8.2).

Combination of Eqs. (8.9) and (8.11) results in an expression for the friction coefficient in binary mixtures, depending on concentration:

$$-\ln \frac{{}^*D_{im}}{RT} = \ln \sigma_{im}\eta_m = x_i \ln {}^*\sigma_{ii}\eta_i + x_j \ln {}^\infty\sigma_{ij}\eta_j \tag{8.12}$$

where ${}^*\sigma_{ii}$ is defined by the equation

$${}^*D_{ii} = \frac{RT}{{}^*\sigma_{ii}\eta_i} \tag{8.13}$$

Since

$${}^\infty D_{ji} = \frac{RT}{{}^\infty\sigma_{ji}\eta_i}$$

Table 10. Basic Self-Diffusivities

Components		V_i/V_j	Diffusivities × 10^5 (cm^2/s) measured		*D$_{ii}$ calculated by equation[a]			
i	j		$^\infty$D$_{ji}$	*D$_{ii}$	1	2	3	4
	Benzene	1.22	1.90					
Cyclohexane				1.49	1.69	1.62	2.09	1.66
	Cyclohexane	0.82	2.10					
Benzene				2.25	2.37	2.45	2.64	2.53
	Acetone	1.22	2.74					
Benzene				2.25	2.43	2.34	3.02	2.27
	Benzene	0.82	4.15					
Acetone				4.85	4.68	4.84	5.22	5.00
	Chloroform	0.90	3.65					
Acetone				4.85	3.89	3.96	4.45	4.03
	Acetone	0.30	1.28					
Water				2.45	2.64	3.08	1.91	2.89

[a] 1, Wilke–Chang; 2, Scheibel; 3, Reddy–Doraiswamy; 4, Lusis–Ratcliff.

it follows that $^*D_{ii}$ is related to $^\infty D_{ji}$ by

$$^*D_{ii} = \frac{^\infty\sigma_{ji}}{^*\sigma_{ii}}\, {}^\infty D_{ji} \tag{8.14}$$

The latter equation enables calculation of the basic self-diffusivity of i from the basic diffusivity of another compound in i if the ratio of the respective size and shape factors is known. The various semiempirical correlations, collected in Table 9, permit estimation of this ratio, which (except for the King–Hsueh–Mao equation) depends only on the (partial) molar volume of the compounds involved.

In Table 10 the basic self-diffusivities estimated from the various correlations and the measured basic diffusivities are compared with measured basic self-diffusivities.[55] Apart from the Reddy–Doraiswamy equation, the calculated data show reasonable agreement with experimental data.

An equation corresponding to Eq. (8.12), but extended to multicomponent mixtures, is obtained by combining Eqs. (8.1), (8.10), and (8.13):

$$-\ln\frac{^*D_{im}}{RT} = \ln(\sigma_{im}\eta_{m}) = x_i \ln(^*\sigma_{ii}\eta_i) + \sum_{\substack{j=1\\ j\neq i}}^{n} x_j \ln(^\infty\sigma_{ij}\eta_j) \tag{8.15}$$

This equation suggests that it is possible to predict diffusivity in multicomponent mixtures from the pure-component viscosities and pure-component and binary "size and shape" factors. The basis of Eq. (8.15) is the Vigne equation, which for equal values of σ_{ij} implies the validity of Andrade's equation (8.5) for mixture viscosity. We will consider this more closely in the next section.

8.4. MIXTURE VISCOSITY

Andrade's equation (8.5) can be written as

$$\frac{x_i \ln(\eta_i/\eta_m)}{x_j \ln(\eta_j/\eta_m)} = -1 \tag{8.16}$$

In Tables 11 and 12 mixture viscosities, calculated with this equation, are compared with measured data. It follows that reasonable agreement is observed only if the molar volume of the components involved are about the same.

Table 11. Comparison of Measured and Calculated Viscosity of Mixtures

Composition (wt fraction)		Viscosity (cP)					
			calculated according to:				
			Andrade, based on:				
		measured	mole fraction	volume fraction	activity	new equation (8.18)	k
Water	Glycerol						
0	1	1306.2	—	—	—	—	1.663
0.05	0.95	498.4	283.8	835.0	388.0	479.1	
0.10	0.90	206.6	96.3	539.8	114.9	208.8	
0.25	0.75	34.17	14.0	155.5	8.91	34.17	
0.50	0.50	5.91	3.18	23.65	2.35	5.72	
0.75	0.25	2.10	1.52	4.42	1.50	1.98	
1	0	0.98	—	—	—	—	
$M=18$	$M=92.1$						
$d=1.00$	$d=1.26$						
Dutrex	HVI						
0	1	147.2	—	—	—	—	1.045
0.10	0.90	129.6	129.8	134.6	—	130.4	
0.25	0.75	110.0	108.9	117.3	—	110.0	
0.50	0.50	85.31	83.99	92.31	—	84.95	
0.75	0.25	67.59	66.94	71.66	—	67.44	
1	0	54.81	—	—	—	—	
$M=350$	$M=460$						
$d=0.983$	$d=0.880$						
Dutrex	Toluene						
0	1	0.589	—	—	—	—	1.079
0.25	0.75	0.985	0.849	1.648	—	0.872	
0.50	0.50	1.660	1.514	4.923	—	1.605	
0.75	0.25	4.74	4.35	15.79	—	4.74	
1	0	54.81	—	—	—	—	
$M=350$	$M=92.1$						
$d=0.983$	$d=0.866$						

For comparison we calculated the mixture viscosities also by applying volume fractions instead of molar fractions in Eq. (8.16). The results in Tables 11 and 12 again show large deviations from the measured data. In fact, the measured data lie between those calculated from the molar fraction and the volume fraction, which suggests that, possibly, activity must be applied. Therefore, we calculated the mixture viscosity for

Table 12. Comparison of Measured and Calculated Viscosity of Mixtures

Composition (wt fraction)		Viscosity (cP)					
			calculated according to:				
			Andrade, based on:				
		measured	mole fraction	volume fraction	activity	new equation (8.18)	k
Dutrex	MEK						
0	1	0.404	—	—	—	—	1.288
0.25	0.75	0.625	0.553	1.179	—	0.601	
0.50	0.50	1.186	0.934	3.787	—	1.130	
0.75	0.25	3.559	2.634	13.52	—	3.559	
1	0	54.81	—	—	—	—	
$M=350$	$M=72.1$						
$d=0.983$	$d=0.807$						
HVI	Toluene						
0	1	0.589	—	—	—	—	1.745
0.25	0.75	1.069	0.832	2.303	—	1.048	
0.50	0.50	2.506	1.479	9.107	—	2.461	
0.75	0.25	9.94	4.676	36.41	—	9.94	
0.90	0.10	38.61	20.52	84.07	—	38.87	
1	0	147.2	—	—	—	—	
$M=460$	$M=92.1$						
$d=0.880$	$d=0.866$						
HVI	MEK						
0	1	0.404	—	—	—	—	2.170
0.25	0.75	0.773	0.541	1.604	—	0.736	
0.50	0.50	2.025	0.897	6.771	—	1.803	
0.75	0.25	7.94	2.662	30.50	—	7.94	
0.90	0.10	30.72	12.74	77.75	—	34.44	
1	0	147.2	—	—	—	—	
$M=460$	$M=72.1$						
$d=0.880$	$d=0.806$						

glycerol–water mixtures using in Eq. (8.16) activity, obtained via group contribution theory (NRTL). The results collected in Table 11 and in Fig. 16 again show insufficient agreement with experiment.

Excellent confirmation is achieved, however, by replacing in Eq. (8.16) the right hand side (-1) by a constant $(-k)$, the value of which is calculated from the measured viscosity of one of the mixtures. It follows from Tables 11 and 12 and from Figs. 16 and 17 that a good fit with

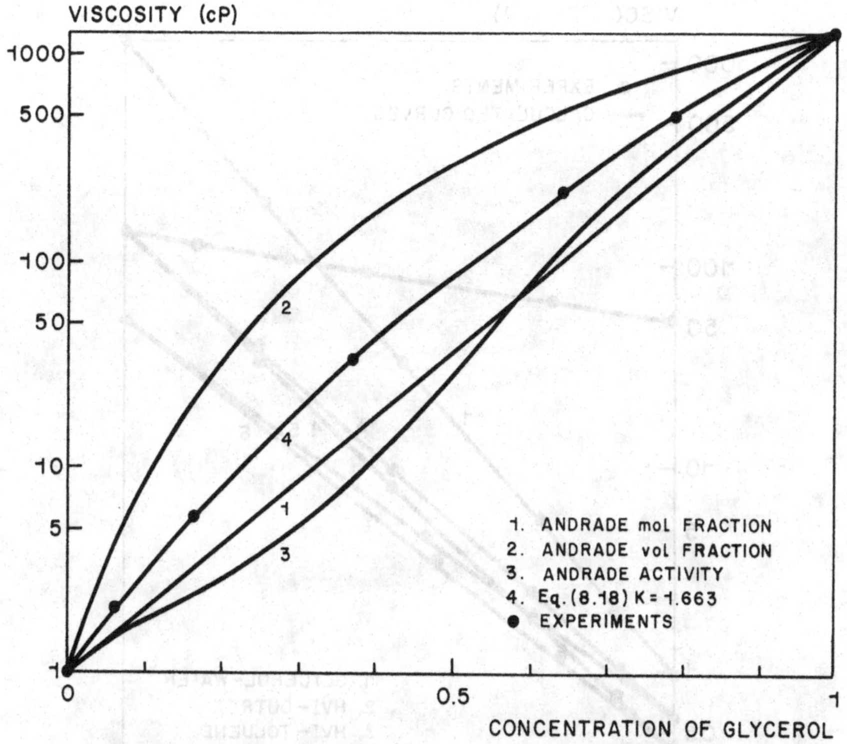

Figure 16. Prediction of the viscosity of glycerol–water mixtures.

experimental results is obtained, even with very extreme systems such as mixtures of lubricating oil and methylethylketone and of water and glycerol.

It is interesting to note that the k values derived for the various mixtures (and also collected in Tables 11 and 12) deviate more from unity if the differences in molar volume and polarity of the compounds involved (hence their nonideality) increases. We believe that the explanation of this phenomenon is as follows. According to Andrade's theory[74] liquid viscosity is mainly due to the existence of associations (pre-structures) in the liquids, disruption of which during forced flow causes the friction force. This model explains, e.g., the decrease in viscosity with increasing temperature observed with liquids, unlike gas viscosity, which increases with temperature in agreement with the kinetic gas theory.

It will be clear that in liquid mixtures the concentration of associations, originally present in the pure components, is affected by the presence of the other components while new, combined associations of the mixture compounds may be formed, depending on their degree of interaction.

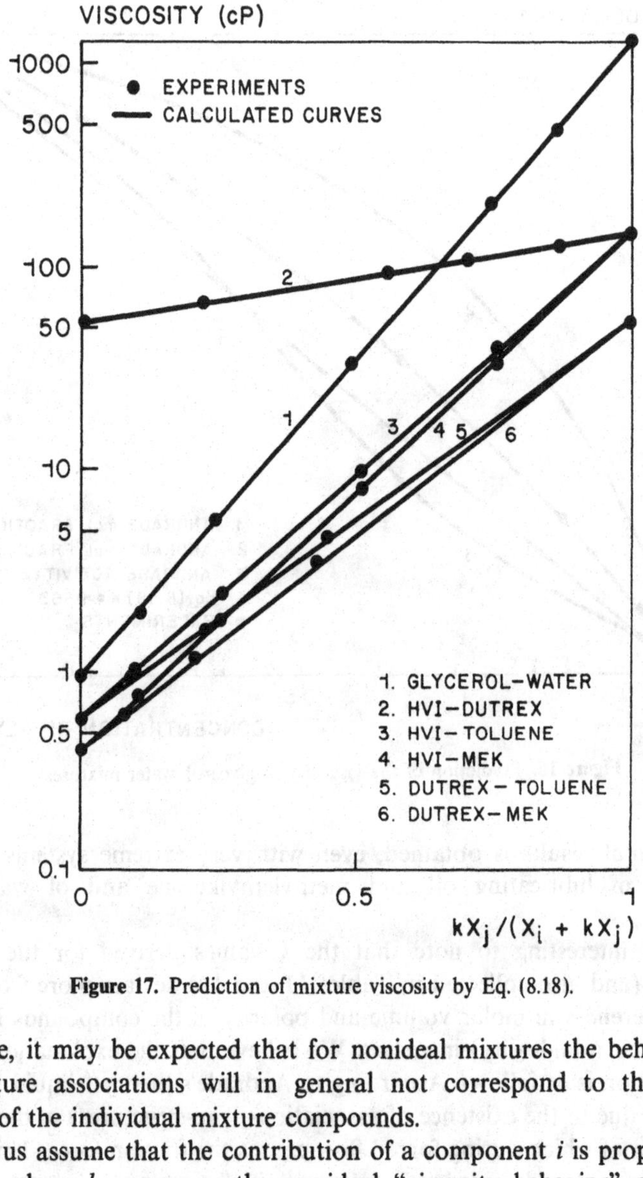

Figure 17. Prediction of mixture viscosity by Eq. (8.18).

Therefore, it may be expected that for nonideal mixtures the behavior of the mixture associations will in general not correspond to the molar average of the individual mixture compounds.

Let us assume that the contribution of a component i is proportional to $k_i x_i$, where k_i expresses the nonideal "viscosity behavior" of i in a mixture. Then the equation of Andrade transforms into

$$(k_i x_i + k_j x_j) \ln \eta_m = k_i x_i \ln \eta_i + k_j x_j \ln \eta_j \tag{8.17}$$

This equation corresponds to Eq. (8.16) after replacing the right-hand side by $-k$, where $k = k_j/k_i$; the equation has been verified experimentally.

In Tables 11 and 12 and in Fig. 17, it is shown that within the experimental error the viscosity of binary mixtures of different concentrations can be described by one single value of k (hence by a constant ratio k_j/k_i), unlike the activity coefficients (Γ_i) and their ratios, which for nonideal systems usually show a strong dependence on concentration.

From the foregoing it is clear that k_i is determined by the associated molecules, while it may be expected that the nonassociated or less-associated molecules contribute mainly to the activity, hence to Γ_i. Consequently, these factors represent quite different parameters. Nevertheless, they both depend on molar interactions and are affected by changing associations due to mixing. Such changes in associations are accompanied by a change in the temperature dependence of these parameters, which in turn is reflected by a heat of mixing. This stresses once more the need of a reliable physical model for prediction of the heat of mixing, being also essential for predicting solubilities of permeants in polymers.

Although the lubricating oil and aromatic oil fraction (Dutrex) consist of a large number of different compounds, their mixture viscosity and that of mixtures with other compounds could be expressed accurately by Eq. (8.17), indicating that this relation is also valid for multicomponent mixtures. In Appendix X the following general equation for the viscosity of multicomponent mixtures is derived:

$$\boxed{\ln \eta_{\mathrm{m}} = \sum_{i=1}^{n} (k_i x_i \ln \eta_i) \Big/ \sum_{i=1}^{n} k_i x_i} \tag{8.18}$$

8.5. PREDICTION OF DIFFUSIVITY FROM MIXTURE VISCOSITY

Of course Eq. (8.18) has consequences for Eq. (8.15), the equation which expresses the self-diffusivity of a component in a multicomponent mixture. The "size and shape" factor of component i in the mixture $\sigma_{i\mathrm{m}}$ is probably also expressed by an equation like

$$\sum_{i=1}^{n} k_i x_i \ln \sigma_{i\mathrm{m}} = k_i x_i \ln {}^{*}\sigma_{ii} + \sum_{\substack{j=1 \\ j \neq i}}^{n} k_j x_j \ln {}^{\infty}\sigma_{ij} \tag{8.19}$$

Even if this statement is not correct, only minor faults are introduced because the differences in σ_{ij} for the various compounds are generally relatively small.

If Eqs. (8.18) and (8.19) are combined, the following modified version of Eq. (8.15) results:

$$\left(\sum_{i=1}^{n} k_i x_i\right) \ln \sigma_{im}\eta_m = k_i x_i \ln {}^*\sigma_{ii}\eta_i + \sum_{\substack{j=1 \\ j\neq i}}^{n} k_j x_j \ln {}^\infty\sigma_{ij}\eta_j \qquad (8.20)$$

When we recall the Stokes–Einstein equation (8.11), it follows that Eq. (8.20) can be written in the form

$$\boxed{\left(\sum_{i=1}^{n} k_i x_i\right) \ln {}^*D_{im} = k_i x_i \ln {}^*D_{ii} + \sum_{\substack{j=1 \\ j\neq i}}^{n} k_j x_j \ln {}^\infty D_{ij}} \qquad (8.21)$$

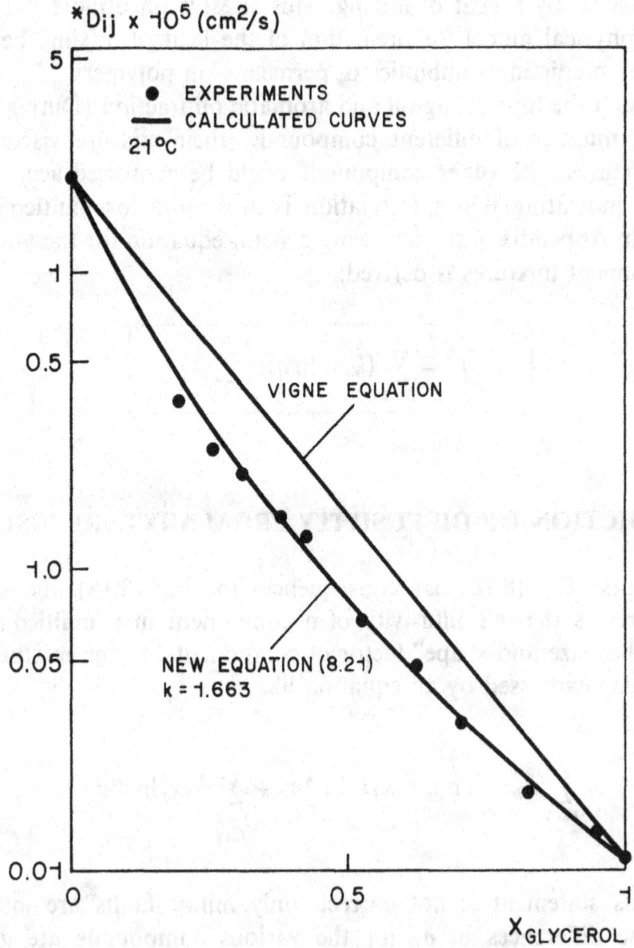

Figure 18. Self-diffusivity of water in glycerol.

This is an interesting result because it suggests that k, which describes the nonideal behavior of mixture viscosity, can be used to calculate the self-diffusion coefficients in mixtures. We checked this for literature data of water–glycerol mixtures at 20°C.[79] Extrapolation of these data yields $1.1 \cdot 10^{-7}$ cm^2/s for the basic diffusivity of water in glycerol. According to Anderson[55] the value of $^*D_{ii}$ for water equals $2.4 \cdot 10^{-5}$ cm^2/s at 25°C, corresponding to about $2.8 \cdot 10^{-5}$ cm^2/s at 20°C.

These basic diffusivities and the value $k = 1.663$, derived from the viscosity data of glycerol–water mixtures in Fig. 16, were applied to obtain the results shown in Fig. 18, which shows indeed excellent agreement between measured and calculated diffusivities.

9

NEW PERMEABILITY EQUATIONS

9.1. INTRODUCTION

The various modes of operation of membrane separation processes are closely related, and can be elucidated as follows: With dialysis and pervaporation the concentration of the permeants on the downstream side of the membrane is kept low by dilution and evacuation, respectively. The resulting concentration gradient in the membrane causes diffusive transport. The transport mechanism seems to be more complicated for pressure-induced permeation (reverse osmosis), but in this case, too, a concentration gradient is created inside the membrane. In the high-pressure side, take-up of permeants occurs due to their thermodynamic potential in the retentate (feed) phase. In the permeate phase the thermodynamic potential of the permeants is lower than in the retentate phase, due to the lower pressure. As a result, the concentration of the permeants in the membrane is lower on the permeate side than on the retentate side, thus causing a concentration gradient inside the membrane.

It can be shown that with thermoosmosis a concentration gradient is likewise formed inside the membrane. Because no consistent theory is available, as yet, for predicting the effects of temperature on free energy and, furthermore, the commercial feasibility of such temperature-driven processes is doubtful from the standpoint of energy consumption,[80] from this point onward we shall consider isothermal processes only.

The extension of the different driving forces for transport, exerted externally in the various membrane processes, can be chosen such that the resulting concentration gradients inside similar membranes are identical. The permeation rates and selectivities in these processes are then also identical. Consequently, all processes can be described by the same permeation equation, if the transport mechanisms inside the membranes are considered instead of those outside.

In our opinion the solution-diffusion theory is most appropriate for the derivation of a membrane permeation equation. This theory applies primarily to tight (diffusion-type) membranes. Transport in such membranes is described by the modified Maxwell–Stefan equation (5.11) which, under steady-state conditions, transforms into Eq. (5.10) of Darken, Prager, and Crank (see Section 5.4). The latter equation for one-dimensional flow reads

$$J_i = -AC_i \frac{{}^*D_{im}}{RT} \frac{\delta \mu_i}{\delta z} \tag{9.1}$$

where A denotes the effective surface area per unit membrane area available for mass transport.

According to Eq. (9.1) the permeate flux depends on permeate concentration and self-diffusion in the membrane material, the parameters of which were considered in Chapters 6 and 8, respectively.

The effective area A is highly influenced by the crystallinity of the polymer; in semicrystalline polymers transport mainly takes place via the amorphous polymer phase, because diffusion through crystallites is neglegibly small.[62–64] Swelling of the polymer in the permeants also has a large influence on A, the effect of which has been ignored in the existing permeation equations.

In the following sections a permeation equation is derived that accounts for the various effects mentioned.

9.2. EFFECT OF CRYSTALLINITY AND SWELLING ON PERMEABILITY

In Section 8.2 we argued that it is advantageous to express the concentration in polymer systems as a volume fraction Φ_i; the permeate flux is then expressed as a volume flow. On substituting in Eq. (9.1) and expressing this equation in integral form, we obtain

$$J_i \int_{z=0}^{l} \frac{\delta z}{A} = -\int_{\mu_{i1}}^{\mu_{i2}} \frac{{}^*D_{im}}{RT} \Phi_i \delta \mu_i \tag{9.2}$$

This equation contains two integrals to be solved: the integral on the left-hand side, which depends on geometry only and is referred to as the geometric diffusion resistance R_{geom}, and the right-hand-side integral, which represents a specific diffusion rate and depends only on physical and thermodynamic parameters.

In a previous paper[4] the following relation was derived for the geometric diffusion resistance:

$$R_{\text{geom}} = R_0 \int_{z=0}^{l} \Phi_{\text{a}}^m \, \delta z \tag{9.3}$$

where R_0 is the geometric diffusion resistance prior to swelling and l is the membrane thickness after swelling. The exponent m equals 2 for semicrystalline polymers and may vary between 0 and $\frac{2}{3}$ for amorphous polymers.

Further, it has been shown (see also Appendix XI) that the geometric diffusion resistance for noncrystalline polymers can be approximated by

$$R_{\text{geom}} = R_0 l_0 (\Phi_{\text{a,av}})^{2n-1} \tag{9.4}$$

and for polymers with a high degree of crystallinity by

$$R_{\text{geom}} = R_0 l_0 (\phi_{\text{a}} + \phi_{\text{c}} \Phi_{\text{a,av}})^{1-n} (\Phi_{\text{a,av}})^{1+n} \tag{9.5}$$

where l_0 is the membrane thickness prior to swelling, ϕ_{a} the volume fraction of amorphous polymer in nonswollen semicrystalline polymer, ϕ_{c} the volume fraction of crystalline polymer in nonswollen semicrystalline polymer, $\Phi_{\text{a,av}}$ the average volume fraction of polymer in amorphous polymer phase, and n is the exponent expressing the degree of anisotropy in swelling.

For noncrystalline polymers $R_0 = 1$, while for semicrystalline polymers

$$R_0 = \frac{M_{\text{c}}^2}{4\pi N \phi_{\text{a}}^2} \tag{9.6}$$

where M_{c} is the surface area of crystallites per unit volume of unswollen polymer and N is the number of "amorphous polymer channels" per unit area of unswollen membrane. Both M_{c} and N are constants if no dissolution of the crystals occurs. Furthermore, we derived that for highly crystalline polymers (such as polyolefins) R_0 can be approximated by

$$R_0 = 3 \frac{\phi_{\text{c}}^{1/3}}{\phi_{\text{a}}} \tag{9.7}$$

In Table 13 various values of R_0, calculated by Eq. (9.7) from the crystallinity of several polyolefin membranes, are compared with those derived

Table 13. Properties of Examined Membranes

Membrane	Thickness (μm)	Temp. (°C)	Density (g/cm^3)	Crystallinity ϕ_c (volume fraction)	Geometric diffusion resistance R_o from experiments	from equations
Polyethylene	10	25	0.9360	0.712	9.09	9.30
Paper coated with polyethylene	26	25	0.9335	0.690	6.1	8.56
Polypropylene	6	55	0.8747	0.430	3.55	3.97

from the permeation measurements with the corresponding membranes. In spite of the rough approximations made in the derivation of Eq. (9.7), reasonable agreement is achieved. Furthermore, the diffusive flux appears to be considerably less in the presence of crystallites. A crystallinity of 70% vol reduces the flux even by a factor of 10; this figure has also been reported by Peterlin.[62]

It is interesting to note that all asymmetric membranes, being prepared by a demixing procedure, consist of semicrystalline polymers, which are obviously inherently less suitable materials for membrane application.

The exponent n in Eqs. (9.4) and (9.5) expresses the degree of anisotropy in swelling. If the membrane is fixed onto a rigid support layer, which prevents extension of the surface area, swelling can lead to an increase in thickness only. In that case $n = 0$ (fully anisotropic swelling). If, however, the membrane material is not supported or reinforced by a non-swelling structure, the membrane is dilated in all dimensions by the same factor and $n = 2/3$ (isotropic swelling).

Equation (9.5) predicts decreasing geometric diffusion resistance with swelling (decreasing $\Phi_{a,av}$) of semicrystalline polymers. This effect is less pronounced with anisotropic swelling. According to Eq. (9.4) R_{geom} of non-crystalline polymers also decreases slightly with isotropic swelling. However, with pronounced anisotropic swelling (small value of n) R_{geom} increases with uptake of permeants due to the overriding effect of increasing membrane thickness.

9.3. SPECIFIC DIFFUSION RATE

The right-hand side of Eq. (9.2) represents the specific diffusion rate, namely, the diffusion rate of a component i through a unit area of unit thickness of amorphous swollen polymer phase. In order to solve the integral on the right-hand side the relationship between the concentration, diffusivity, and the gradient of the thermodynamic potential of the permeants must be known.

The extended and modified Vigne equations (8.10) and (8.21), respectively, enable one to calculate the self-diffusivity of a component from its basic self-diffusivity and basic diffusivities in the compounds involved, and their concentrations.

The concentration of the various permeants can be derived from their thermodynamic potential by means of the modified Flory–Huggins equation (6.18). Consequently, the required relationships for calculating the specific diffusion rates are available. However, because of their complexity, these equations must be solved numerically across the membrane, starting, e.g., on the upstream side and applying an iteration procedure until the calculated downstream conditions coincide with the actual ones.

This procedure is very laborious and its complexity increases dramatically with the number of permeants. We therefore developed a semi-empirical procedure[4] to solve the right-hand side of Eq. (9.2).

9.4. SIMPLIFIED CALCULATION PROCEDURE OF MEMBRANE PERMEATION

During the numerical calculations on binary permeant systems we observed that all concentration-dependent parameters (q) vary as a function of position z inside the membrane according to more or less the same function $f(z)$:

$$q(z) = q_2 - (q_2 - q_1)\, f(z) \tag{9.8}$$

in which 1 and 2 denote upstream side ($z = 0$) and downstream side ($z = 1$), respectively.

The boundary values of $f(z)$ are: $f(0) = 1$ and $f(1) = 0$. If the gradient in the thermodynamic potential is given by

$$d\mu_i = dF_i = RT d \ln a_i + V_i\, dP - S_i\, dT \tag{9.9}$$

where a_i is the activity of component i and S_i the partial molar entropy of component i, then for isothermal processes Eq. (9.2) can be transformed into

$$J_i R_{\mathrm{geom}} = -\int_{f(z)=0}^{1} {}^*D_{im} \Phi_i \left(d \ln a_i + \frac{V_i}{RT} dP \right) \tag{9.10}$$

In order to solve this equation the pressure distribution inside the membrane must be known. This will be examined in the next section.

9.5. STRESS DISTRIBUTIONS INSIDE MEMBRANES

Rosenbaum and Cotton[81] studied possible pressure distributions inside membranes by examining reverse osmosis of aqueous salt solutions through supported membranes of cellulose diacetate. In interpreting their results they considered three different cases: systems with a constant pressure inside the membrane equal to the upstream pressure and to the downstream pressure, and porous membranes showing viscous flow resulting in a linear pressure gradient. They observed the existence of a concentration gradient of water inside the membrane that increased with increasing permeation pressure. According to the authors their results support the theory of diffusive transport and, furthermore, makes plausible a constant (upstream) pressure inside the membrane.

Indeed, existing theories do not account for a pressure drop inside the membranes. It is assumed that the pressure drop occurs at the interface of the membrane and the porous support. It is argued by Paul *et al.*[82] that mass transport then takes place by a concentration gradient, caused by the lower thermodynamic potential of the permeants at the downstream boundary as a result of the lower pressure. Quantitatively, for isothermal and incompressible systems at equilibrium

$$ {}_{\mathrm{m}}a_{i2} = {}_{l}a_{i2} \exp\left[-\frac{V_i}{RT}(P_1 - P_2) \right] \tag{9.11} $$

in which m and l denote the membrane and fluid phase, respectively.

If the membrane consists of a swollen *noncrystalline polymer* (such as an elastomer) supported by a mechanically nondeforming construction, the applied permeation pressure is taken up by the support. In that case the pressure inside the main part of the membrane is constant and equals that of the upstream side. Only near the pores of the support is there a pressure distribution, which probably has a rather complicated shape and is difficult to predict.

With *semicrystalline polymers* different cases may be distinguished. If the crystallites are surrounded by amorphous polymer (which is generally true[83]) both phases are loaded to the pressure exerted if the membrane is supported, so that this case resembles the former one.

With *highly crystalline materials* it might be possible that the pressure exerted is taken up by the crystallites if they form a continuous phase across the membrane thickness. Except for a thin layer on the upstream side the amorphous polymer is then not loaded. This is only possible, however, if the pressure drop in that layer does not exceed its compressive strength. Otherwise deformation takes place and may eventually result in loading of the total amorphous phase to upstream pressure.

The pressure distribution inside composite membranes is difficult to predict due to the presence of an intermediate layer of deviating mechanical properties and a porous substrate. It is often assumed that in self-supporting membranes, such as homogeneous hollow fibers, a pressure gradient also occurs inside the membrane due to the permeation pressure applied. It can be shown, however, that for noncompressible materials (Poisson's ratio 0.5, which applies to most polymers) the hydrostatic stress σ_h (which in fact equals the average of the radial, tangential, and axial stresses) is the same everywhere and equal to[84]

$$\sigma_h = (P_1 R_1^2 - P_2 R_2^2)/(R_2^2 - R_1^2) \tag{9.12}$$

where R is the radius (see Fig. 19). Hence the force exerted on the permeants is the same throughout the membrane.

If $R_1 \approx R_2 = R$, then Eq. (9.12) transforms into

$$\sigma_h = -R(P_1 - P_2)/2l \tag{9.13}$$

where l is the membrane thickness. According to this equation the hydrostatic stress is negative (compression) if the excess pressure is exerted on the outside of the fiber, and positive (tension) if it is exerted on the inside. Consequently, different effects can be expected depending on whether pressure-induced permeation takes place from the outside or the inside of the fiber.

Furthermore, the hydrostatic stress increases with increasing fiber diameter and decreasing wall thickness.

As the pressure exerted on the permeants inside the membrane material depends on the hydrostatic stress, both phenomena influence the activity (concentration) gradients inside the membrane as expressed by Eq. (9.11). By considering homogeneous flat membranes supported by an open screen spacer, like, e.g., in spiral-wound modules, about the same stress conditions prevail as in hollow fibers loaded on the inside (see Fig. 19) and, consequently, an almost constant hydrostatic stress inside the membrane can also be expected. Compressive stress concentrations may be expected at the contact places of the membrane with the spacer, however, that might effect the transport process, especially if a fine screen spacer is used.

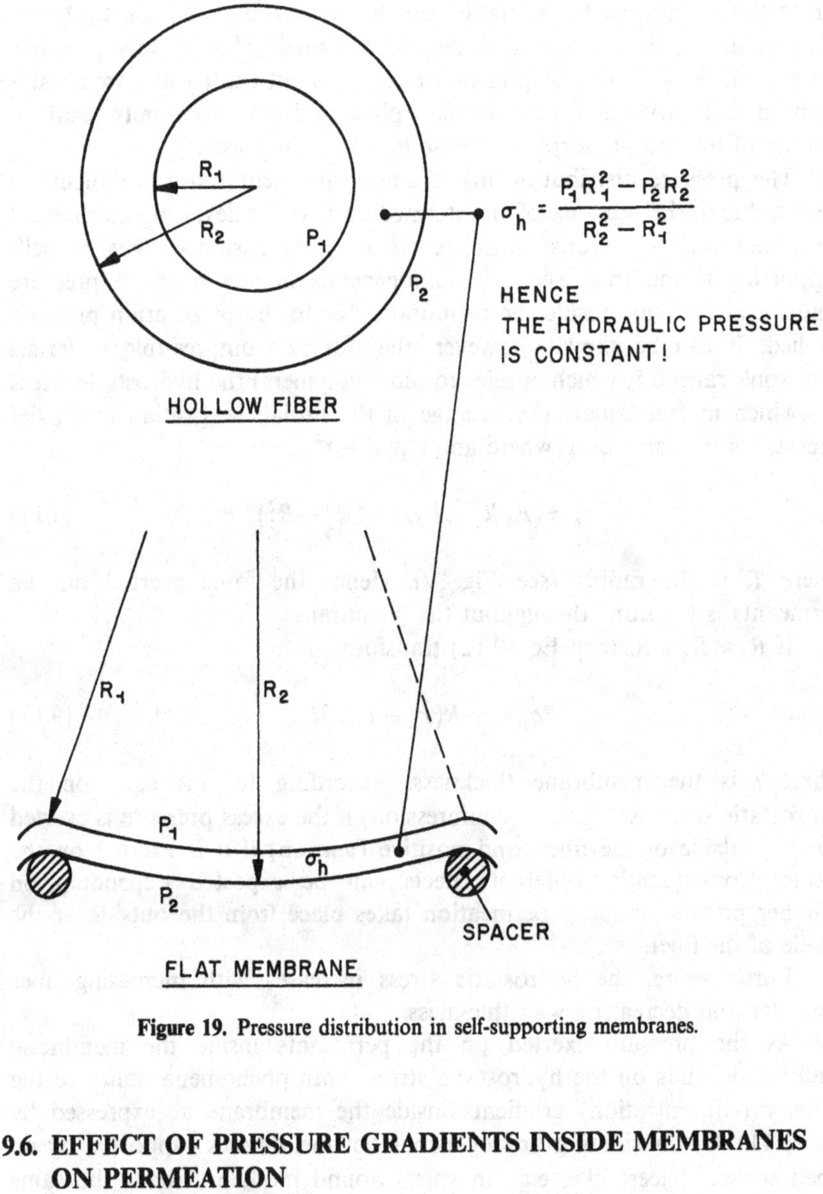

Figure 19. Pressure distribution in self-supporting membranes.

9.6. EFFECT OF PRESSURE GRADIENTS INSIDE MEMBRANES ON PERMEATION

An estimate of the effect of different shapes of pressure gradients inside membranes on permeation rate was made by van der Waal.[85] For reverse osmosis of aqueous solutions he calculated a reduction in permeate flux of less than 10% in the most unfavorable case and at the maximum permeation pressure applied in normal practice (100 bar). This relatively small effect is valid only for permeants of low molar volume, such as water. For hydrocarbons, having generally much larger molar volumes, the effect under corresponding conditions is much higher and certainly not negligible, as will be shown below.

If exponential functions are introduced for the pressure and concentration gradients inside the membrane (with exponents β_p and β_c, respectively), then the following general permeation equation (derived in Appendix XII) is obtained by combining Eqs. (9.8) and (9.10):

$$J_i R_{geom} = {}_aD_{im} \cdot {}_a\Phi_i \ln \frac{a_{i1}}{a_{i2}} + {}_a\Phi_i \cdot \Delta{}^*D_{im} + {}^*D_{im,av} \cdot \Delta\Phi_i$$

$$+ \frac{V_i}{RT} \Delta P \left[{}^*D_{im1} \Phi_{i1} - \frac{\beta_c}{\beta_p + \beta_c} (\Delta{}^*D_{im} \Phi_{i2} + {}^*D_{im2}\, \Delta\Phi_i) - \frac{2\beta_c}{\beta_p + 2\beta_c} (\Delta{}^*D_{im}\, \Delta\Phi_i) \right] \tag{9.14}$$

where β_c is the exponent of the exponential concentration gradient in the membrane, β_p the exponent of the exponential pressure gradient inside the membrane,

$${}_aD_i = \frac{a_{i1}{}^*D_{im2} - a_{i2}{}^*D_{im1}}{a_{i1} - a_{i2}}, \qquad {}_a\Phi_i = \frac{a_{i1}\Phi_{i2} - a_{i2}\Phi_{i1}}{a_{i1} - a_{i2}}$$

$$\Delta{}^*D_{im} = {}^*D_{im,1} - {}^*D_{im,2}, \qquad {}^*D_{im,av} = ({}^*D_{im,1} + {}^*D_{im,2})/2$$

$$\Delta\Phi_i = \Phi_{i1} - \Phi_{i2}, \qquad \Delta P = {}_mP_1 - {}_mP_2$$

According to Eq. (9.14), which applies for all isothermal membrane separation processes, the flux J_i depends only on the conditions and diffusivities in the *membrane boundary layers* and the exponents β_c and β_p. In Appendix XI it is shown that β_c can also be derived from the boundary conditions of the membrane; β_p is considered below.

If equilibrium is assumed at the interfaces of the membrane with the retentate and permeate phases, the concentrations of the permeants in the membrane boundary layers can be calculated from the corresponding local thermodynamic potentials using the modified Flory–Huggins equation. From these concentrations the required self-diffusivities can be calculated by means of the modified Vigne equation.

Equation (9.14) enables one to estimate the effect of the pressure distribution inside the membrane on the flux. We consider three cases:

1. The exponent of the *concentration* gradient β_c is zero, which means that the permeant concentration across the membrane is constant and corresponds to the equilibrium concentration under the upstream conditions. The contribution of the pressure-dependent term in Eq. (9.14) then becomes $V_i\,\Delta P\,{}^*D_{im1}\Phi_{i1}/RT$.
2. The exponent of the *pressure* gradient β_p is zero, which means that the pressure inside the membrane is constant and equals the upstream pressure. Then the pressure-dependent term transforms into $V_i\,\Delta P\,{}^*D_{im2}\Phi_{i2}/RT$.
3. The exponents of the concentration gradient and pressure gradient are the same (e.g., $\beta_c = \beta_p = 1$, meaning linear gradients). The pressure-dependent term becomes

$$V_i\,\Delta P[({}^*D_{im1}\Phi_{i1} + {}^*D_{im2}\Phi_{i2})/2 - \Delta{}^*D_{im}\,\Delta\Phi_i/6]/RT$$

As both the diffusivity and concentration are higher on the upstream side, it follows that the contribution of the pressure-dependent term to the flux is maximum in the first case.

Let us assume that the permeant concentration on the downstream side is negligibly small (hence Φ_{i2} and a_{i2} vanish); then the pressure-independent term of Eq. (9.14) becomes equal to ${}^*D_{im1}\Phi_{i1}$, and the flux becomes

$$J_i R_{\text{geom}} = {}^*D_{im1}\Phi_{i1}(1 + V_i\,\Delta P/RT)$$

According to this equation the flux of a component with a molar volume of 100 ml, at room temperature and 100 bar permeation pressure (which is set equal to the pressure drop inside the membrane), is increased by 40% due to the pressure effect. The flux becomes even larger with larger molar volumes of the permeants.

It must be emphasized, however, that the case considered here represents an ultimate effect under conditions which can probably never be fulfilled, such as $\beta_c = 0$, $a_{i2}(\Phi_{i2}) = 0$. As pointed out in Section 9.5, there is in general no pressure gradient inside a membrane so that Eq. (9.14) reduces to

$$J_i R_{\text{geom}} = {}_{\text{a}}D_{im} \cdot {}_{\text{a}}\Phi_i \ln\frac{a_{i1}}{a_{i2}} + {}_{\text{a}}\Phi_i \cdot \Delta{}^*D_{im} + {}^*D_{im,\text{av}} \cdot \Delta\Phi_i \qquad (9.15)$$

9.7. PERMEABILITY EQUATION FOR GASES

We argued previously that gases dissolved in a polymer must be regarded as being in the liquid phase, so Eqs. (9.14) and (9.15) are valid for liquids as well as gases. The many physical and thermodynamic parameters required in these equations are usually unavailable for gases. It was shown in Section 6.6, however, that for the solubility of gases in polymers the modified Flory–Huggins equation can be reduced to

$$C_i = \frac{{}^{\infty}H_i f_i}{(1-\Phi_s)} \exp(B_i \Phi_s) \tag{6.23}$$

In this equation the concentration C_i, expressed in cm^3 (STP)/cm^3, can be substituted by the volume fraction Φ_i with the aid of Eq. (6.24):

$$\Phi_i = \frac{V_i}{\phi_a} C_i \Big/ \left(c + \frac{V_i}{\phi_a} C_i \right) = \frac{V_i}{\phi_a} \frac{C_i}{c} \Phi_a$$

This approach yields

$$\Phi_i = \left(\frac{V_i {}^{\infty}H_i}{\phi_a c} \right) f_i \exp(B_i \Phi_i) \tag{9.16}$$

The activity of a gas is expressed by Eq. (6.20),

$$a_i = f_i / f_{sat,i}$$

hence the fugacity in Eq. (9.16) can be replaced to give the equation

$$\Phi_i = \left(\frac{V_i {}^{\infty}H_i}{\phi_a c} f_{sat,i} \right) a_i \exp(B_i \Phi_s)$$

If we take the logarithm of this equation and differentiate we obtain

$$d \ln a_i = d \ln \Phi_i - B_i \, d\Phi_s \tag{9.17}$$

since the term within the large parentheses on the right-hand side of the previous equation is constant.

With Eq. (9.17) the integral in the permeability equation (9.10) can be solved, omitting the pressure-dependent term and using Eq. (8.10) to calculate the self-diffusivity in mixtures. In this way the next permeability equation for gases is derived in Appendix XIII:

$$J_i R_{\text{geom}} = {}^*D_{im,\ln}\,\Delta\Phi_i + \frac{B_i}{\theta}\,{}^*D_{im,\ln}\,\Delta\Phi_i\,\Delta\Phi_s - \frac{B_i}{\theta}\left({}^*D_{im1}\Phi_{i1} - {}^*D_{im2}\Phi_{i2}\right)\Delta\Phi_s \tag{9.18}$$

where

$${}^*D_{im,\ln} = \frac{{}^*D_{im1} - {}^*D_{im2}}{\ln({}^*D_{im1}/{}^*D_{im2})} \quad \text{and} \quad \theta = \ln({}^*D_{im1}/{}^*D_{im2})$$

10

PERMEATION EXPERIMENTS (PROGRAM AND PROCEDURES)

It was pointed out in Section 9.1 that the various membrane processes, such as pervaporation, reverse osmosis, dialysis, and gas permeation, are closely related and must therefore in principle be described by the same theoretical model. The former three processes concern liquid separations, for which we derived the general permeation rate equation (9.15). In order to check this equation for the processes mentioned permeation experiments were performed with hydrocarbons through polyolefin membranes. Furthermore, literature data for the permeation of aqueous ethanol solutions through cellulose diacetate were used to check the validity of Eq. (9.15) for strongly polar systems.

Although Eq. (9.15) is also valid for the separation of gases, we derived another equation, Eq. (9.18), in which a simplified form of the modified Flory–Huggins equation for the solubility of the gases has been applied, thus reducing the number of required parameters. We checked this equation, using literature data for the permeation of CO_2 and CH_4 through polysulfone, depending on temperature and permeation pressure.[44]

In the sections below we discuss the various experiments and results, starting with the liquid separations.

10.1. MEMBRANES TESTED

The permeation experiments with hydrocarbons were carried out with membranes of (commercial) homogeneous, polyethylene and polypropylene sheets. Some physical data and properties of the membranes are given in Tables 13 and 14. The densities of the polymers were determined

Table 14. Thermodynamic and Physical Data

Compound	Temp. (°C)	Molar volume (cm^3) at T	Molar volume (cm^3) at boiling point	Density of amorphous polymer (g/cm^3)	Heat of evaporation at boiling point (cal/mol)	Viscosity (cP)	Solubility parameter $\left(\frac{cal}{cm^3}\right)^{1/2}$	$\frac{L_{av}-s_o}{s_o}\cdot\frac{y}{y-1}$
Polyethylene sheet	25	16.37[a]	—	0.8555	—	—	8.41	6.92
Polyethylene (coated paper)	25	16.37	—	0.8555	—	—	8.41	7.23
Polypropylene (400% stretched)	25	16.35	—	0.8565	—	—	8.00	9.99
Polypropylene	55	16.72	—	0.8372	—	—	7.77	9.99
n-Hexane	55	137.6	140.1	0.625	6896	0.2261	6.55	—
n-Heptane	25	147.0	164.8	0.680	7576	0.3867	7.45	—
	55	153.4	164.8	0.652	7576	0.2840	6.79	—
n-Octane	55	168.9	188.8	0.675	8225	0.3650	7.20	—
2,2,4-Trimethylpentane	25	166.0	179.2	0.687	7411	0.4699	6.85	—
	55	171.1	179.2	0.666	7411	0.3378	6.53	—
n-Hexadecane	55	301.3	350	0.750	12240	1.715	7.68	—

[a] Volume of a CH_2 group.

with a pycnometer at the temperature of the permeation tests. The crystallinities of the polymers were calculated from their densities, as described in Appendix VIII.

10.2. PERVAPORATION EXPERIMENTS

The pervaporation experiments were performed with the apparatus shown in Fig. 20. The membrane (a 10 μm polyethylene sheet) was sup-

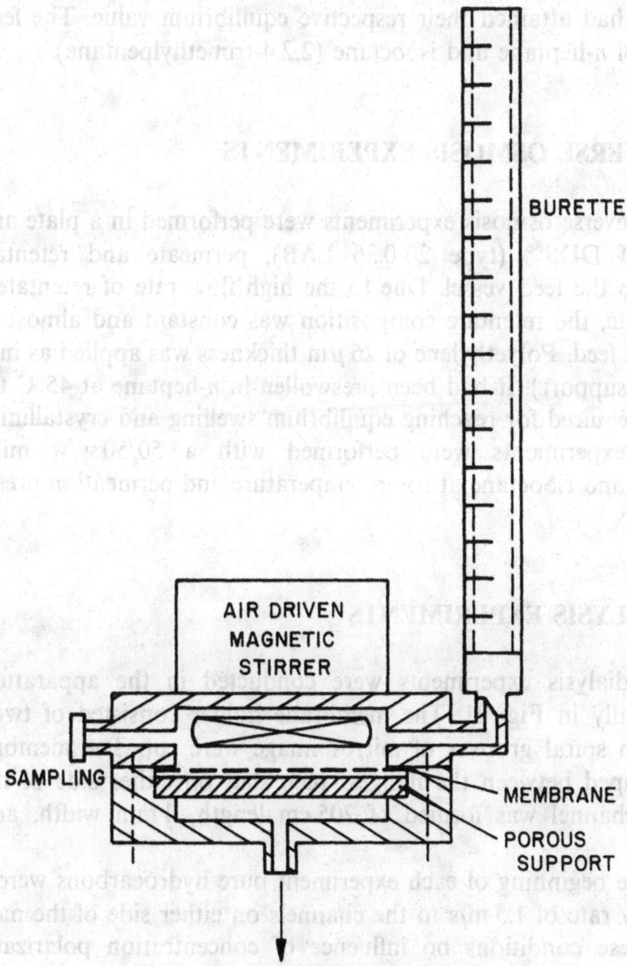

Figure 20. Pervaporation apparatus.

ported by a porous metal plate. The retentate phase was mixed vigorously with an air-driven magnetic stirrer. The feed was supplied via a burette to determine the flux. The downstream side of the membrane holder was evacuated and the permeate vapors were condensed in a cold trap.

Reliable results were obtained after reaching steady-state conditions, i.e., when the composition of the permeate had become equal to the feed composition in the burette. Samples of permeate and retentate were then taken and analyzed by GLC.

The duration of the experiments was at least 10 days in order to ensure that polymer swelling and crystallinity and membrane flux and selectivity had attained their respective equilibrium value. The feeds were mixtures of *n*-heptane and isooctane (2,2,4-trimethylpentane).

10.3. REVERSE OSMOSIS EXPERIMENTS

The reverse osmosis experiments were performed in a plate and frame module of DDS[86] (type 20-0.36 LAB); permeate and retentate were recycled to the feed vessel. Due to the high flow rate of retentate relative to permeate, the retentate composition was constant and almost equal to that of the feed. Polyethylene of 26 μm thickness was applied as membrane (paper as support). It had been preswollen in *n*-heptane at 45°C to reduce the time required for reaching equilibrium swelling and crystallinity.

The experiments were performed with a 50/50 w/w mixture of *n*-heptane and isooctane at room temperature and permeation pressures up to 75 bar.

10.4. DIALYSIS EXPERIMENTS

The dialysis experiments were conducted in the apparatus shown schematically in Fig. 21. The membrane holder consisted of two flanges into which spiral grooves of mirror image were cut. The membrane had been clamped between the flanges, such that on either side of the membrane a channel was formed of 205 cm length, 4 mm width, and 2 mm depth.

At the beginning of each experiment pure hydrocarbons were fed at a linear flow rate of 1.5 m/s to the channels on either side of the membrane. Under these conditions no influence of concentration polarization was observed.

In order to limit the pressure difference across the fragile membrane both loops were connected to storage vessels held under atmospheric con-

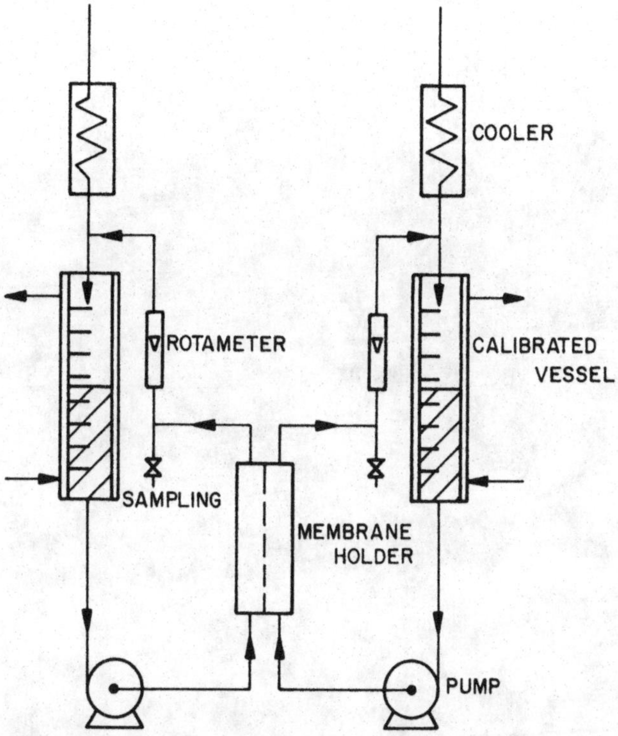

Figure 21. Dialysis apparatus.

ditions. Net mass flow across the membrane was read from the calibrated storage vessels and the composition of liquid in either loop was determined by GLC, which then allowed calculation of the flux of the individual hydrocarbons.

The dialysis experiments were performed at 55°C with various model compounds ($n-C_6$, $n-C_7$, $n-C_8$, $i-C_8$, $n-C_{16}$) and a 6 μm polypropylene sheet as membrane.

11

RESULTS OF PERMEATION EXPERIMENTS

11.1. EFFECT OF PERMEATION TIME ON FLUX AND SELECTIVITY

The measured fluxes and selectivities in various pervaporation and reverse osmosis experiments have been plotted versus permeation time in Figs. 22 and 23. The fluxes increased with time probably due to swelling and partial dissolution of crystallites in the polymer. Without pretreatment, more than a week was required to reach steady-state conditions at room temperature (see Fig. 22). However, on preswelling this time was reduced dramatically as is shown in Fig. 23 for reverse osmosis. Here preswelling was carried out at elevated temperature (45°C) in *n*-heptane.

The selectivity α_j^i, which is defined by

$$\alpha_j^i = \frac{J_i}{J_j} \bigg/ \frac{{}_l\Phi_{i1}}{{}_l\Phi_{j1}} \tag{11.1}$$

where ${}_l\Phi_{i1}$ is the volume fraction of component i in the retentate, was calculated from the permeate and retentate compositions. Generally the selectivity decreased with permeation time, again probably due to swelling.

In the reverse osmosis experiments with preswollen membranes, however, selectivity increased with time (Fig. 23). Recent evidence indicates that this effect was caused by the formation of submicron pores during the preswelling state. Most of them disappeared with time, possibly due to reorientation of the polymer chains.

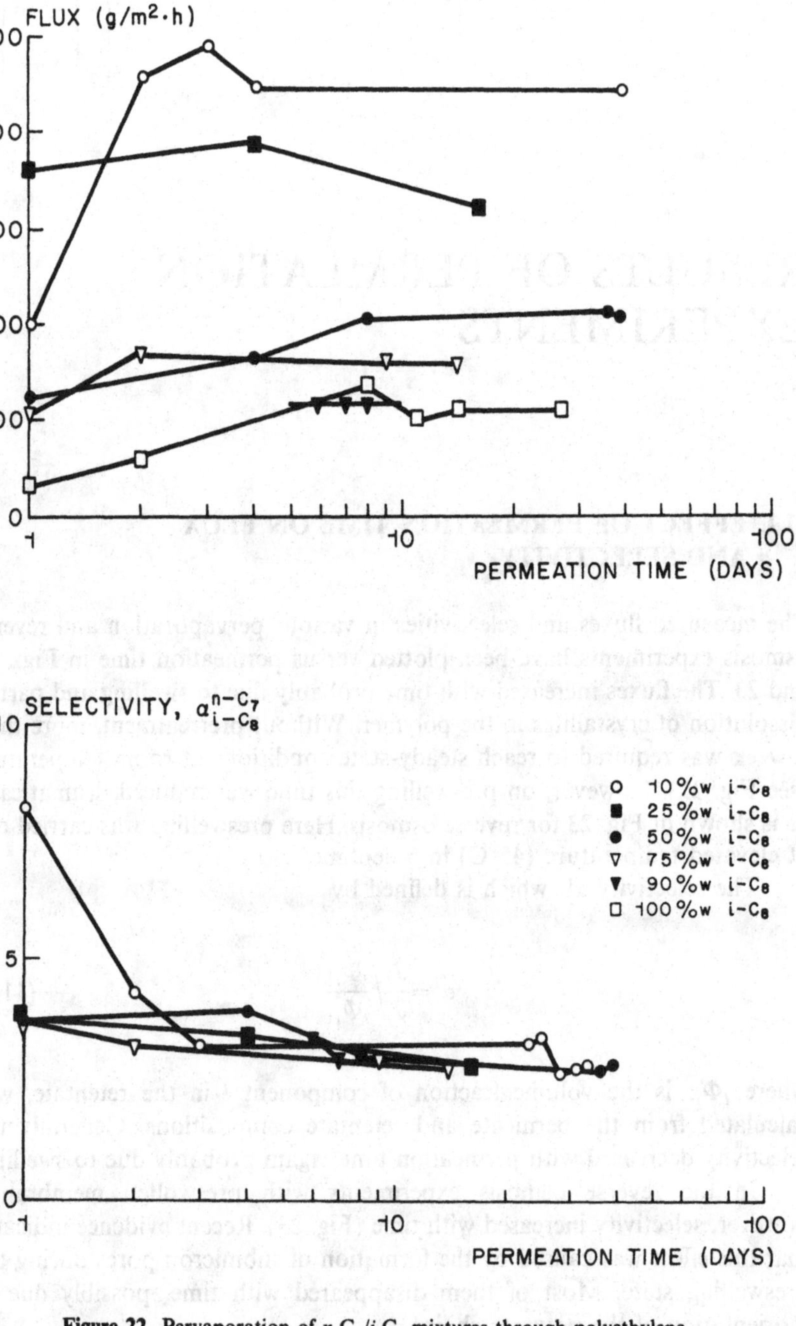

Figure 22. Pervaporation of n-C_7/i-C_8 mixtures through polyethylene.

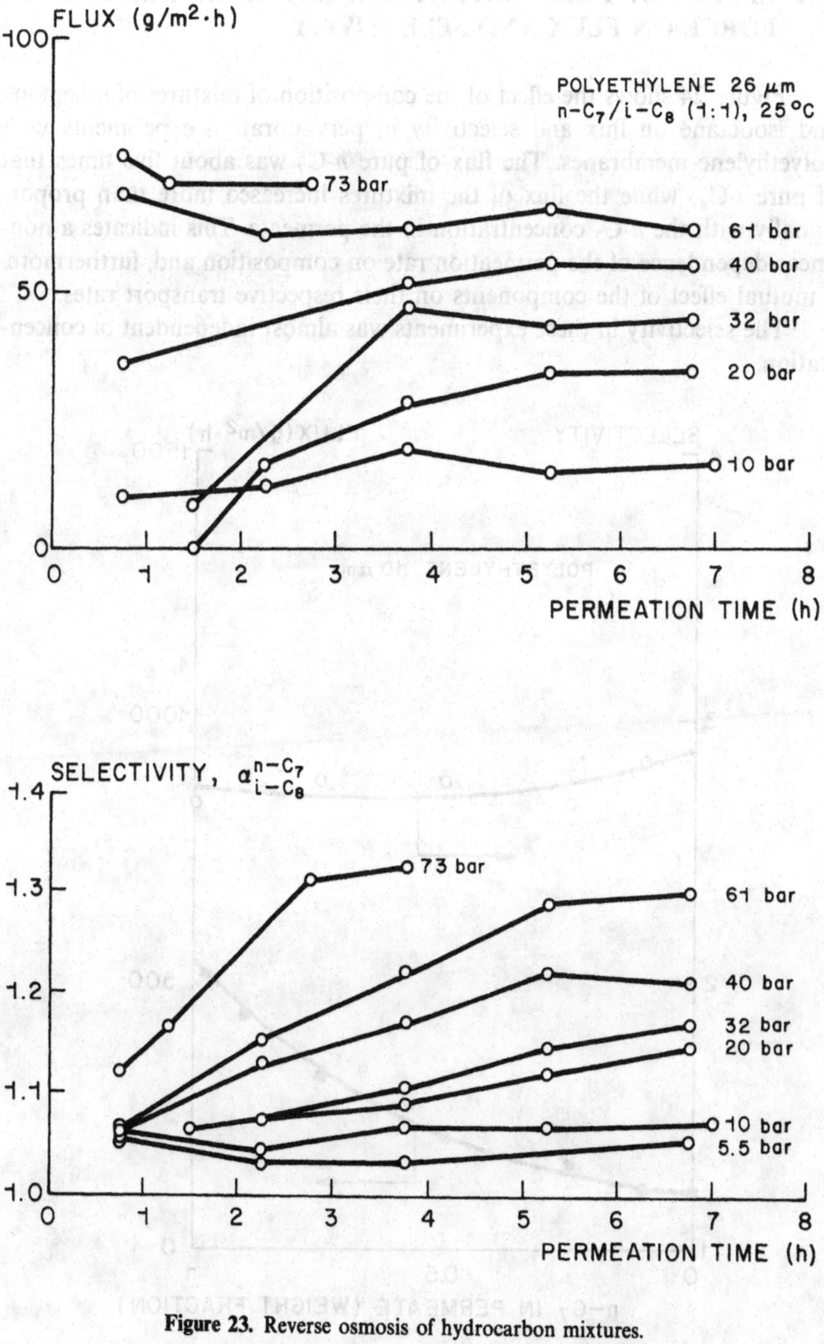

Figure 23. Reverse osmosis of hydrocarbon mixtures.

11.2. EFFECT OF COMPOSITION AND EXTERNAL DRIVING FORCE ON FLUX AND SELECTIVITY

Figure 24 shows the effect of the composition of mixtures of *n*-heptane and isooctane on flux and selectivity in pervaporation experiments with polyethylene membranes. The flux of pure *n*-C_7 was about five times that of pure *i*-C_8, while the flux of the mixtures increased more than proportionally with the *n*-C_7 concentration in the permeate. This indicates a nonlinear dependence of the permeation rate on composition and, furthermore, a mutual effect of the components on their respective transport rates.

The selectivity in these experiments was almost independent of concentration.

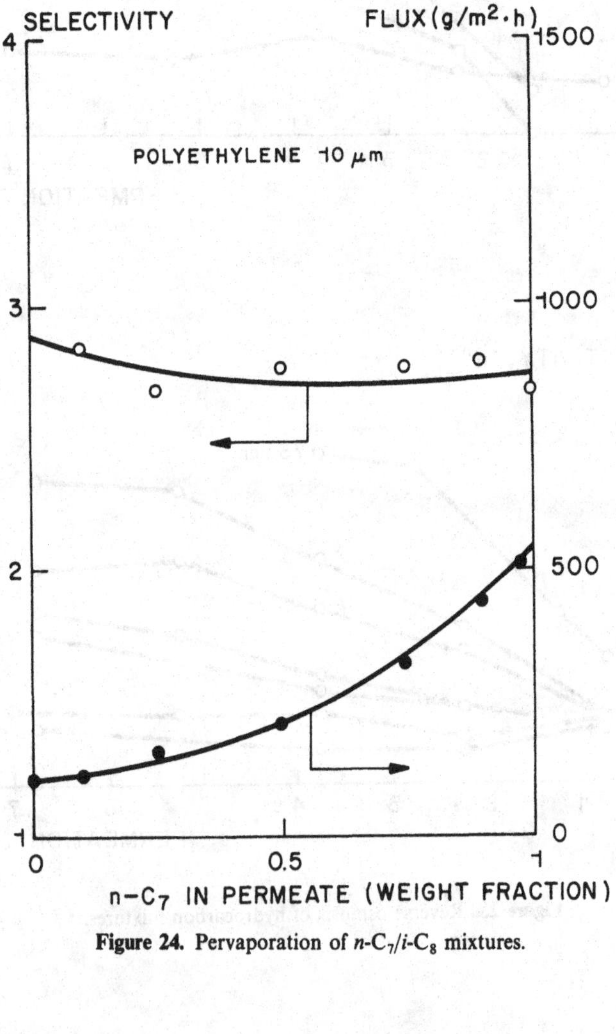

Figure 24. Pervaporation of *n*-C_7/*i*-C_8 mixtures.

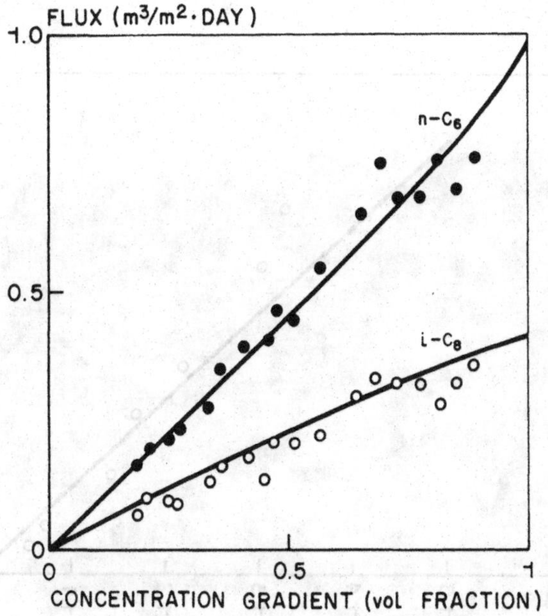

Figure 25. Dialysis experiments: —— calculated curves, ∘ experiments, 6 μm polypropylene (55 °C).

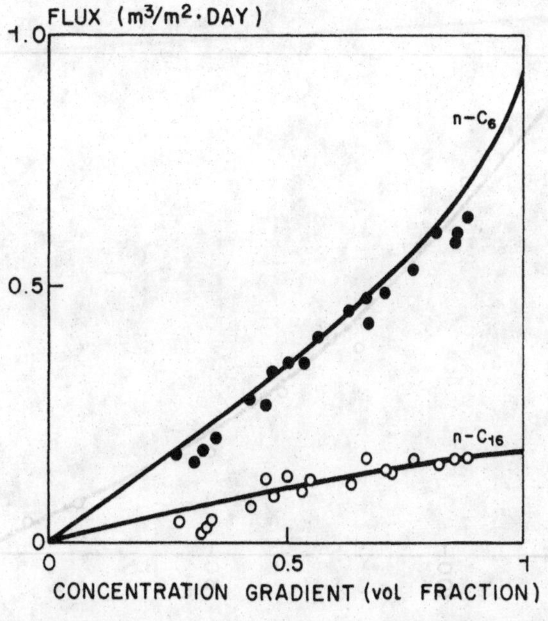

Figure 26. Dialysis experiments: —— calculated curves, ∘ experiments, 6 μm polypropylene (55 °C).

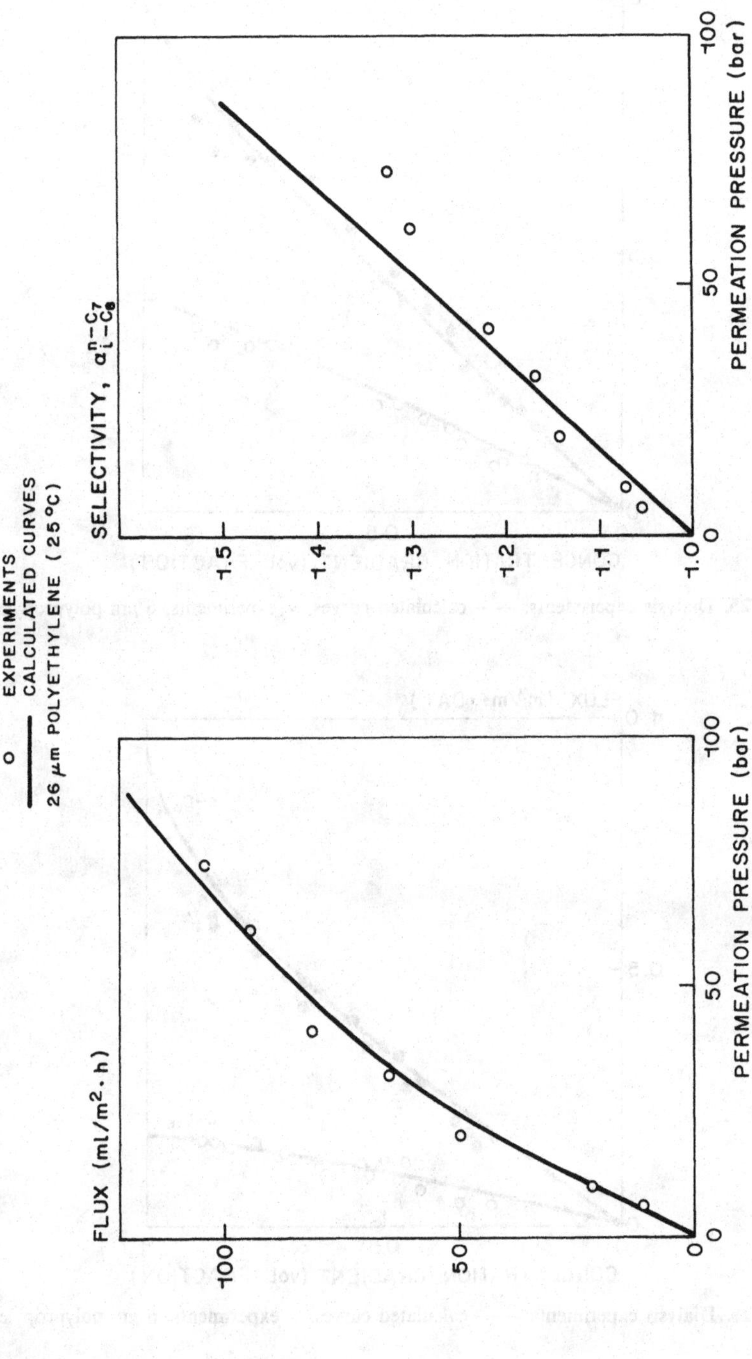

Figure 27. Reverse osmosis of n-C_7/i-C_8 mixture (1:1 w).

The results of some dialysis experiments with polypropylene membranes are shown in Figs. 25 and 26. The flux increased almost proportionally with the concentration gradient of the hydrocarbons across the membrane and was much larger than that observed in the pervaporation experiments with polyethylene membranes. The selectivity was lower.

Apart from the different membrane materials, these effects are due to the higher temperature applied in the dialysis experiments and to the larger degree of swelling.

Figure 27 represents the flux and selectivity of reverse osmosis experiments with a 50/50 w/w mixture of *n*-heptane and isooctane through a polyethylene membrane as a function of permeation pressure. The flux levels off with increasing permeation pressure, in accordance with earlier observations of Paul[82] with rubber membranes. The flux is low compared to dialysis and pervaporation, and so is the selectivity. This can be expected from the relatively low permeation pressures applied in reverse osmosis (maximum 70 bar). According to theory the flux and selectivity of reverse osmosis coincide with those of pervaporation at infinitely large permeation pressure.[3]

12

EXPERIMENTAL CHECK OF THE PERMEATION EQUATIONS

12.1. CONCENTRATION DEPENDENCE OF THE MEAN DIFFUSIVITY

For pervaporation with zero pressure on the permeate side, and if we assume constant pressure inside the membrane (in which case $dP = 0$, $a_{i2} = 0$, and $\Phi_{i2} = 0$), Eq. (9.15) is valid and reduces to

$$J_i R_{\mathrm{geom}} = {}^*D_{im,\mathrm{av}} \Phi_{i1}$$

in which the mean coefficient of self-diffusion inside the membrane is given by

$${}^*D_{im,\mathrm{av}} = ({}^*D_{im,1} + {}^*D_{im,2})/2$$

In Appendix XI we derive that for semicrystalline polymers

$$R_{\mathrm{geom}} = R_0 l \Phi^2_{\mathrm{a,av}}$$

which, combined with the former equation, yields

$$J_i R_0 l \Phi^2_{\mathrm{a,av}} = {}^*D_{im,\mathrm{av}} \Phi_{i1} \tag{12.1}$$

According to the extended Vigne equation (8.10) the self-diffusivity ${}^*D_{im}$ for a two-component permeant mixture in a polymer satisfies

$$\ln {}^*D_{im} = \Phi_i \ln {}^*D_{ii} + \Phi_j \ln {}^{\infty}D_{ij} + \Phi_{\mathrm{a}} \ln {}^{\infty}D_{ia} \tag{12.2}$$

if Φ_a ($\Phi_{a,av}$) represents the (average) volume fraction of polymer in the *amorphous polymer phase.*

This equation permits calculation of the self-diffusivity in a mixture from the composition of that mixture and the basic self-diffusivity and basic diffusivities of the respective components. The permeant concentrations Φ_i for the various experiments were calculated from the data in Tables 13 and 14, which were derived in the way described in Appendix VIII, applying the modified Flory–Huggins equation (6.18). The thickness l of the swollen membrane and R_0 were calculated according to the procedure given in Appendix XI. The basic self-diffusivities and the binary mutual basic diffusivities of the hydrocarbons (see Table 15) were obtained from the literature, or derived from existing correlations as described in Sections 8.1 and 8.3.

The data thus obtained enabled calculation of the basic diffusivities of the various hydrocarbons in polyolefins from pervaporation experiments (which are also collected in Table 15), using Eq. (12.1). Hence we calculated the mean coefficients of self-diffusion of *n*-heptane and isooctane mixtures in polyethylene. These values are compared in Table 16 and Fig. 28 with the data derived from the corresponding pervaporation experiments; they show reasonable agreement.

Table 15. Basic Diffusivities and Basic Self-Diffusivities

Component		Temp.	Diffusivity ($10^{-7} \times cm^2/s$)			
i	j	(°C)	$^*D_{ii}$	$^\infty D_{ij}$	$^*D_{jj}$	$^\infty D_{ji}$
n-C_6	n-C_7	55	630.3	547.3	509.4	592.7
n-C_6	n-C_8	55	630.3	453.8	399.8	556.1
n-C_6	i-C_8	55	630.3	461.5	427.4	591.0
n-C_6	n-C_{16}	55	630.3	130.6	86.0	411.3
n-C_7	n-C_8	55	509.4	421.4	399.8	476.8
n-C_7	i-C_8	25	340.1	285.4	282.8	337.0
n-C_7	i-C_8	55	509.4	428.5	427.4	506.7
Ethanol	Water	17	50.0	100	200.0	100.0
n-C_6	Polypropylene	55	630.3	17.44	—	—
n-C_7	Polypropylene	55	509.4	8.105	—	—
n-C_8	Polypropylene	55	399.8	6.66	—	—
i-C_8	Polypropylene	55	427.4	7	—	—
n-C_{16}	Polypropylene	55	86.0	2.2	—	—
n-C_7	Polyethylene	25	340.1	2.717	—	—
i-C_8	Polyethylene	25	282.8	1.547	—	—
Ethanol	Cellulose diacetate	17	50.0	0.301	—	—
Water	Cellulose diacetate	17	200.0	0.605	—	—

Table 16. Activities and Swelling Data of Pervaporation Experiment with Polyethylene at 25°C

n-C_7 in permeate (% w)	Flux ($cm^3/m^2 \cdot h$)		Retentate phase: concentration (volume fraction)		Retentate phase: activity		Equilibrium swelling at upstream side (volume fraction)		
	n-C_7	i-C_8	n-C_7	i-C_8	n-C_7	i-C_8	n-C_7	i-C_8	total
0	0	160.2	0	1	0	1	0	0.1394	0.1394
10	17.6	157.3	0.039	0.961	0.0474	0.9653	0.0094	0.1356	0.1450
25	58.8	174.8	0.109	0.891	0.1294	0.8799	0.0264	0.1283	0.1547
50	150.7	149.3	0.268	0.732	0.3058	0.7125	0.0679	0.1079	0.1758
75	358.4	118.3	0.523	0.477	0.5640	0.4484	0.1326	0.0728	0.2054
90	588.7	64.8	0.767	0.233	0.7916	0.2242	0.2041	0.0357	0.2398
98	734.7	14.9	0.947	0.053	0.9532	0.0513	0.2485	0.0080	0.2565
100	—	—	1	0	1	0	0.2622	—	0.2622

n-C_7 in permeate (% w)	β_c	l/l_0	Average self-diffusivity (10^{-7} cm^2/s): n-C_7 measured	n-C_7 calculated	i-C_8 measured	i-C_8 calculated
0	0.605	1.009	—	—	1.64	—
10	0.600	1.010	2.66	2.71	1.64	1.63
25	0.592	1.010	3.11	2.79	1.90	1.69
50	0.575	1.012	3.00	3.00	1.87	1.83
75	0.555	1.014	3.50	3.34	2.10	2.07
90	0.534	1.018	3.54	3.79	2.23	2.39
98	0.525	1.019	3.53	4.05	2.22	2.58
100	0.522	1.020	—	—	—	—

12.2. PREDICTION OF SELECTIVITY IN PERVAPORATION

Combination of Eqs. (11.1) and (12.1) yields the following expression for the selectivity in pervaporation:

$$\alpha_{ij} = \frac{{}^*D_{im,\mathrm{av}}}{{}^*D_{jm,\mathrm{av}}} \frac{\Phi_{i1}}{\Phi_{j1}} \frac{{}_l\Phi_{j1}}{{}_l\Phi_{i1}} \tag{12.3}$$

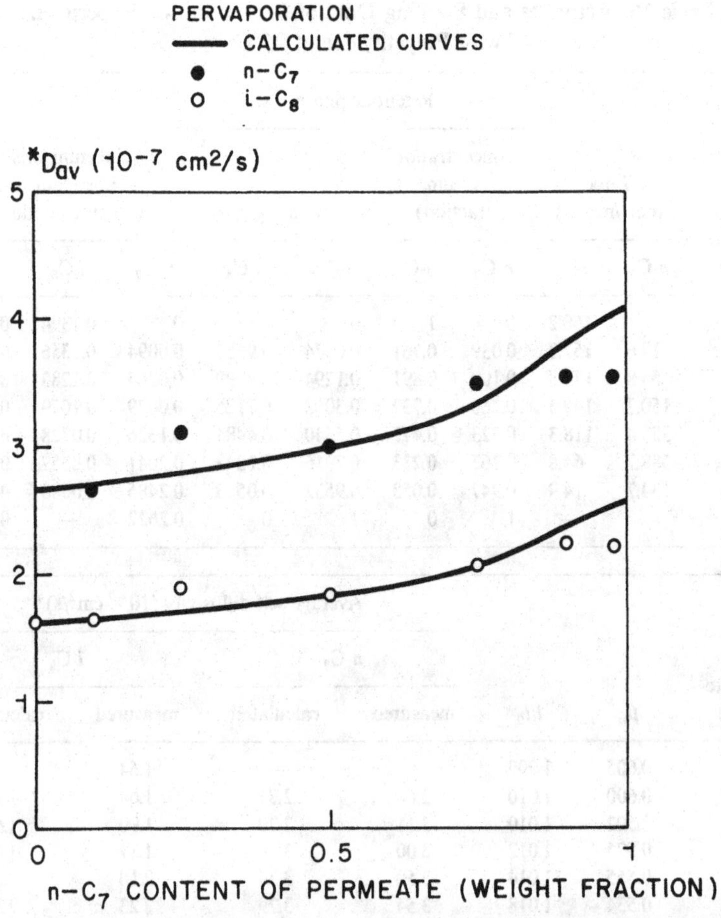

Figure 28. Self-diffusivity of n-C_7 and i-C_8 in polyethylene.

Assuming equilibrium between the retentate and the membrane at its upstream interface, this equation can be transformed into

$$\alpha'^{i}_{j} = \frac{{}^*D_{im,av}}{{}^*D_{jm,av}} \frac{\Gamma_{j1}}{\Gamma_{i1}} \frac{{}_{l}\Gamma_{i1}}{{}_{l}\Gamma_{j1}} \tag{12.4}$$

where ${}_{l}\Gamma_{i1}$ and Γ_{i1} are the activity coefficients based on the volume fraction of component i in the retentate and the upstream side of the membrane, respectively.

Table 17. Comparison of Calculated and Measured Selectivities in Pervaporation Experiments

		Activity coefficient (volume base)					
	n-C_7 content of permeate	retentate phase		upstream side of polymer		Selectivity α_j^i	
Membrane	(% w)	n-C_7	i-C_8	n-C_7	i-C_8	calculated	measured
Polyethylene	0	—	1	—	7.174	—	—
	10	1.2140	0.9951	5.043	7.119	2.87	2.84
	25	1.1904	0.9872	4.902	6.858	2.79	2.68
	50	1.1419	0.9731	4.502	6.603	2.82	2.76
	75	1.0782	0.9402	4.253	6.159	2.68	2.76
	90	1.0321	0.9622	3.878	6.280	2.75	2.78
	98	1.0063	0.9703	3.835	6.413	2.72	2.68
	100	1	—	3.814	—	—	—

Table 17 shows the various activity coefficients calculated from the thermodynamic and physical data contained in Tables 14 and 16. From these data and the calculated mean self-diffusivities we calculated the selectivities according to Eq. (12.4). The results, which are compared in Table 17 with those derived from experiments, show good agreement.

12.3. COMPARISON OF CALCULATED AND MEASURED PERMEATION IN REVERSE OSMOSIS

The basic diffusivities derived from pervaporation experiments with n-C_7/i-C_8 mixtures through polyethylene enable us to calculate the flux of these two individual compounds during reverse osmosis. For this we apply the general permeation equation (9.15), assuming constant (upstream) pressure inside the membrane. The results are presented in Table 18 and Fig. 27. It follows from this figure that the calculated flux is in excellent agreement with measurements.

The calculated selectivities, too, are in reasonable agreement with experiments, but tend to deviate somewhat at high permeation pressures. As mentioned before (see Section 11.1), this must probably be attributed to the presence of submicron pores, formed in the membrane during preswelling and causing some leakage, especially at high pressures.

Table 18. Fluxes for Reverse Osmosis Calculated from Basic Diffusivities of 26 μm Polyethylene Coated Paper at 25°C[a]

Selectivity a_{C_7/C_8}	Permeate composition: concentration (volume fraction) n-C_7	i-C_8	Permeate composition: activity n-C_7	i-C_8	Permeation pressure (bar)	Equilibrium swelling at dowstream side: concentration (volume fraction) n-C_7	i-C_8	Equilibrium swelling at dowstream side: activity n-C_7	i-C_8
1.1	0.5629	0.4371	0.6021	0.4198	17.7	0.1229	0.0574	0.5413	0.3723
1.2	0.5842	0.4158	0.6223	0.3993	35.2	0.1080	0.0450	0.5037	0.3144
1.3	0.6035	0.3965	0.6406	0.3807	51.7	0.0964	0.0362	0.4695	0.2680
1.5	0.6372	0.3628	0.6722	0.3477	86.0	0.0770	0.0241	0.4008	0.1943

[a] Composition in feed (volume fraction): n-C_7 = 0.5393, i-C_8 = 0.4607
Activity in feed: n-C_7 = 0.5796, i-C_8 = 0.4427
Equilibrium swelling at upstream side: n-C_7 = 0.1419, i-C_8 = 0.0755

Permeation pressure (bar)	Effective self-diffusivity (10^{-7} cm²/s): upstream side n-C_7	i-C_8	downstream side n-C_7	i-C_8	Calculated flux (total) (cm³/m²·h)
17.7	7.622	4.872	6.566	3.925	41.1
35.2	7.622	4.872	5.864	3.312	70.1
51.7	7.622	4.872	5.374	2.895	92.0
86.0	7.622	4.872	4.664	2.330	104.0

12.4. COMPARISON OF CALCULATED AND MEASURED PERMEATION BY DIALYSIS

In Fig. 29 the composition of the n-C_6-rich phase is plotted versus its concentration drop across the membrane in the dialysis experiments where n-C_6 was part of a mixture with either isooctane or n-C_{16}. Under these conditions we calculated first the self-diffusivities inside the membrane by means of Eq. (12.2) and then the respective fluxes (see Table 19) by means of Eq. (9.15). The calculated curves in Figs. 25 and 26 coincide nicely with experimental data points.

Table 19. Fluxes for Dialysis Calculated from Basic Diffusivities of 6 μm Polypropylene at 55°C

Concentration gradient (volume fraction)	n-C_6 in feed (volume fraction)	Activities at				Equilibrium swelling at			
		upstream side		downstream side		upstream side		downstream side	
		n-C_6	i-C_8	n-C_6	i-C_8	n-C_6	i-C_8	n-C_6	i-C_8
1	1	1	0	0	1	0.2997	0	0	0.2515
0.926	0.982	0.9855	0.0144	0.0680	0.9313	0.2945	0.0044	0.0173	0.2372
0.736	0.913	0.9288	0.0704	0.2095	0.7886	0.2744	0.0213	0.0545	0.2061
0.560	0.837	0.8644	0.1342	0.3209	0.6766	0.2520	0.0399	0.0851	0.1804
0.339	0.722	0.7648	0.2330	0.4339	0.5632	0.2188	0.0679	0.1170	0.1535
0.222	0.655	0.7016	0.2959	0.4855	0.5116	0.1984	0.0851	0.1321	0.1407
		n-C_6	n-C_{16}	n-C_6	n-C_{16}	n-C_6	n-C_{16}	n-C_6	n-C_{16}
1	1	1	0	0	1	0.2997	0	0	0.2837
0.886	0.990	0.9954	0.0066	0.2259	0.7987	0.2978	0.0028	0.0487	0.2483
0.674	0.919	0.9625	0.0524	0.4527	0.5919	0.2854	0.0220	0.1070	0.2036
0.480	0.836	0.9225	0.1047	0.5855	0.4665	0.2698	0.0437	0.1469	0.1706
0.320	0.759	0.8830	0.1534	0.6658	0.3878	0.2542	0.0635	0.1734	0.1470

Concentration gradient (volume fraction)	l/l_0	Effective self-diffusivity (10^{-7} cm^2/s)				Calculated flux (m^3/m$^2\cdot$day)	
		upstream side		downstream side			
		n-C_6	i-C_8	n-C_6	i-C_8	n-C_6	i-C_8
1	1.068	51.02	26.45	39.68	19.69	0.984	0.420
0.926	1.068	50.81	26.32	40.29	20.04	0.889	0.379
0.736	1.069	49.96	25.81	41.58	20.80	0.686	0.326
0.560	1.069	49.03	25.22	42.66	21.44	0.514	0.253
0.339	1.069	47.68	24.43	43.80	22.11	0.308	0.156
0.222	1.069	46.88	23.95	—	—	—	—
		n-C_6	n-C_{16}	n-C_6	n-C_{16}	n-C_6	n-C_{16}
1	1.073	51.02	10.55	28.89	6.22	0.939	0.187
0.886	1.075	50.96	10.55	34.18	7.05	0.723	0.174
0.674	1.079	50.67	10.61	38.51	8.12	0.503	0.136
0.480	1.081	50.05	10.59	41.58	8.87	0.343	0.108
0.320	1.082	49.25	10.50	43.61	9.34	0.225	0.074

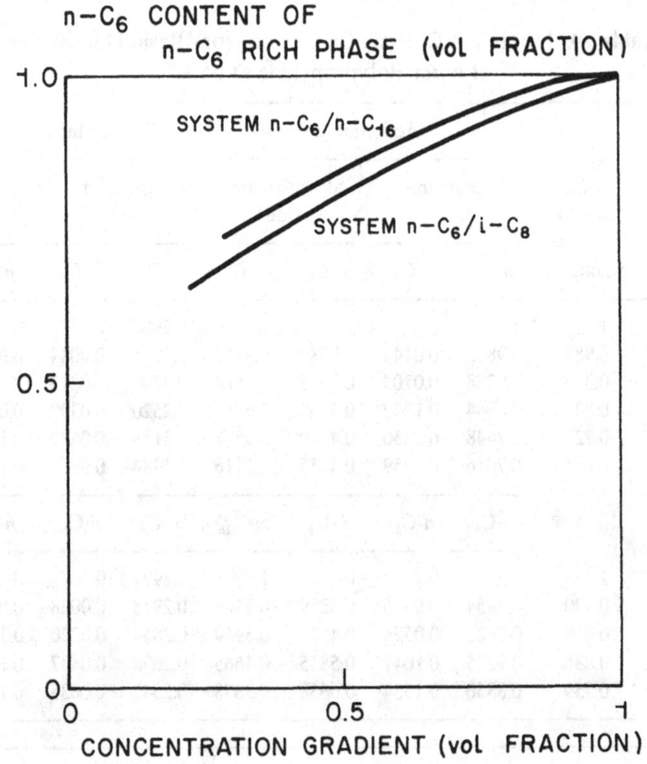

Figure 29. Experimental conditions of dialysis experiments.

12.5. MUTUAL EFFECT OF HYDROCARBONS ON THEIR PERMEABILITY

It is well known that the flux of individual components is affected by the presence of others; this effect is called coupled transport. Contrary to the conventional permeation equations, which account for this effect by incorporation of an arbitrary coupling coefficient (see also Sections 3.1 and 3.4), Eq. (9.15) uses meaningful physical parameters only.

By way of example, Fig. 30 shows curves calculated by means of this equation from the available physical and thermodynamic data. It follows that the flux of n-C_6 in dialysis increases with increasing flux of the counter solvent applied, being in quantitative agreement with the experimental results shown also in this figure.

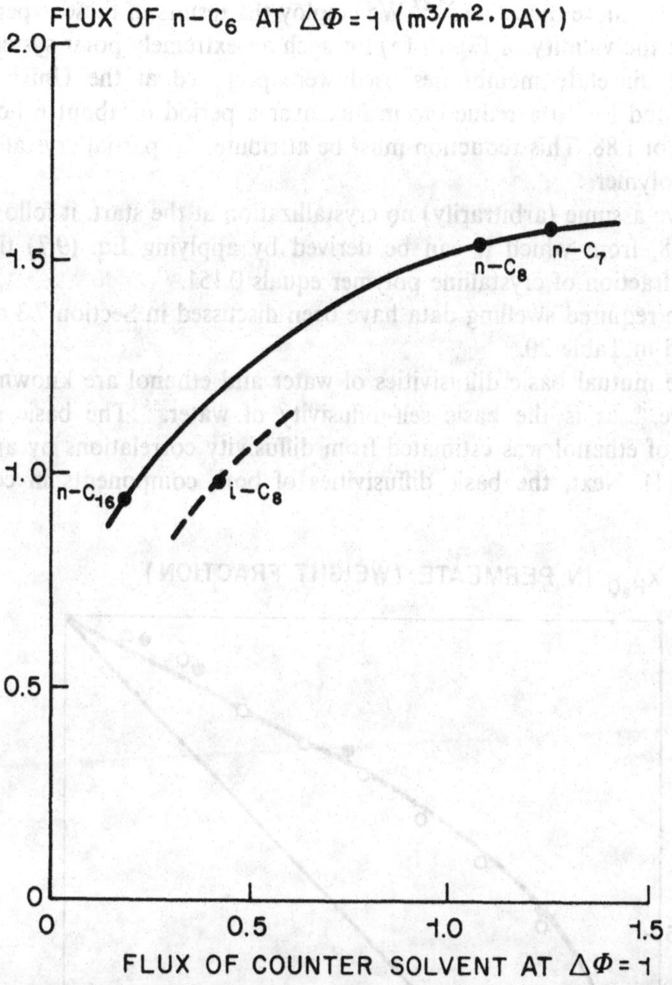

Figure 30. Coupled transport of hydrocarbons.

12.6. PERVAPORATION OF ETHYL ALCOHOL/WATER MIXTURES THROUGH CELLULOSE DIACETATE MEMBRANES

The systems considered in the previous sections all consist of nonpolar components, which show only weak dispersion interactions.

Extensive pervaporation studies were performed at the University of Twente (The Netherlands) with mixtures of ethyl alcohol and water through cellulose diacetate.[53,87] We employ the results of these experiments to check the validity of Eq. (9.15) for such an extremely polar system. The cellulose diacetate membranes used were prepared at the University of Twente and led to a reduction in flux over a period of about 6 hours by a factor of 1.88. This reduction must be attributed to partial crystallization of the polymer.

If we assume (arbitrarily) no crystallization at the start, it follows that $R_0 = 1.88$, from which it can be derived by applying Eq. (9.7) that the volume fraction of crystalline polymer equals 0.151.

The required swelling data have been discussed in Section 7.3 and are collected in Table 20.

The mutual basic diffusivities of water and ethanol are known in the literature,[56] as is the basic self-diffusivity of water.[55] The basic self-diffusivity of ethanol was estimated from diffusivity correlations by applying Eq. (8.14). Next, the basic diffusivities of both components in cellulose

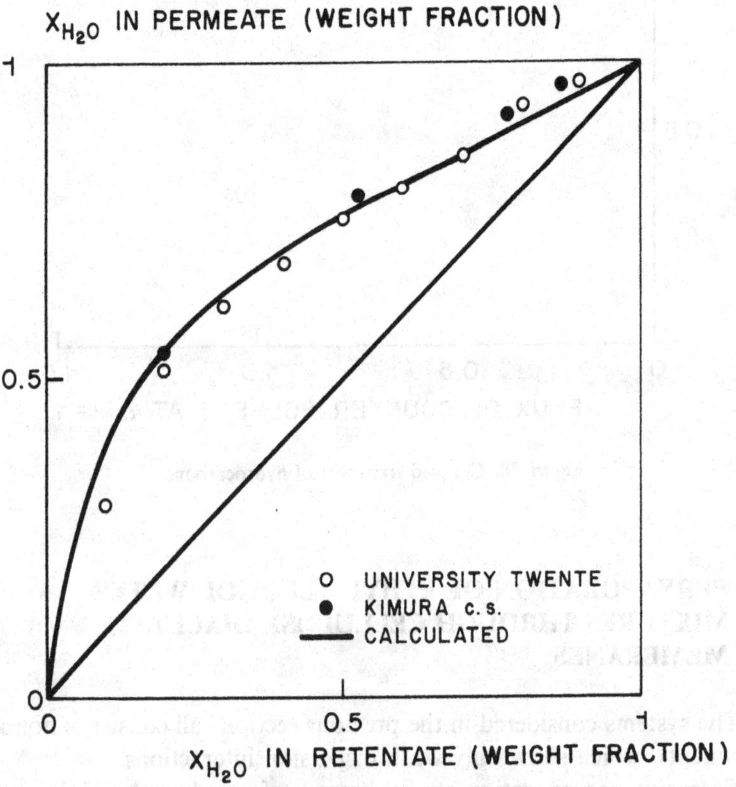

Figure 31. Membrane separation of ErOH/H_2O mixtures.

Table 20. Separation of EtOH/H_2O Mixtures with Cellulose Diacetate

Feed H_2O content (weight fraction)	Swelling at upstream		Permeate H_2O content (weight fraction)
	EtOH	H_2O	
1	0	0.1910	1.0000
0.9	0.0264	0.1836	0.9502
0.7	0.0787	0.1703	0.8554
0.5	0.1287	0.1463	0.7562
0.35	0.1664	0.1236	0.6693
0.3	0.1715	0.1085	0.6331
0.1	0.1941	0.0459	0.3929
0.04	0.1931	0.0199	0.2204
0	0.194	0.0000	0.0000

diacetate were derived from the pure-water and pure-ethanol pervaporation experiments. The various data thus obtained are summarized in Table 15. They were used to calculate the composition of permeate depending on the retentate (feed) composition, and we compared the results with measured pervaporation data of Mulder[88] and Kimura and Nomura.[89] The results in Fig. 31 show excellent confirmation, indicating that the new permeability equation (9.15) is indeed capable of describing quantitatively membrane separation of polar systems.

12.7. PERMEABILITY OF GASES IN POLYSULFONE

Sada *et al.*[44] have measured the solubility and permeability of pure CO_2 and CH_4 in polysulfone (a glassy polymer) as a function of temperature and permeation pressure. The sorption parameters for the dual-mode sorption model as well as for the modified Flory–Huggins model (adapted for gases) are collected in Table 8, from which it follows that fewer parameters are required with the latter model in order to achieve excellent confirmation with experiments, as shown in Fig. 12.

For calculating the permeability Sada *et al.* applied a model, developed by Barrer,[43] based on the dual-mode sorption model, and accounting for diffusion from the bulk polymer phase to the microvoids (which are assumed to be present in the polymer) and *vice versa.* As discussed in Section 7.4 and shown in Table 21, three diffusivity coefficients are required for each permeant apart from the three sorption coefficients.

Table 21. Diffusivities of CO_2 and CH_4 in Polysulfone[44]

		Diffusion parameters according to:				
		dual-mode sortion			modified Flory–Huggins equation	
Gas	Temp. (°C)	D_{DD} (10^{-9} cm²/s)	D_{DH} (10^{-9} cm²/s)	$(D_{HH}+D_{HD})$ (10^{-9} cm²/s)	${}^{\infty}D_{ip}/R_{geom}$ (10^{-7} cm/s)	$\ln(D_{ii}/{}^{\infty}D_{ip})$
CO_2	30	39.5	26.8	4.53	48.58	8.113
	35	51.2	29.1	5.55	64.20	8.383
	40	62.8	30.4	7.25	73.83	8.859
	45	75.1	33.9	8.85	89.87	8.955
CH_4	30	1.85	0.945	0.428	4.275	7.352
	35	2.18	1.21	0.595	5.728	6.184
	40	2.66	1.51	0.931	7.977	5.305
	45	3.21	1.99	1.38	8.395	7.191

The same experimental data are employed to check our permeation rate equation (9.18) for gases:

$$J_i R_{geom} = {}^*D_{im,ln}\, \Delta\Phi_i + \frac{B_i}{\theta}\, {}^*D_{im,ln}\, \Delta\Phi_i\, \Delta\Phi_s - \frac{B_i}{\theta}\,({}^*D_{im,1}\Phi_{i1} - {}^*D_{im,2}\Phi_{i2})\, \Delta\Phi_s$$

For a single permeant this equation transforms into (see Appendix XIII)

$$J_i R_{geom} = {}^{\infty}D_{ip} \left\{ \frac{[\exp(\Theta\Phi_{i1}) - \exp(\Theta\Phi_{i2})]}{\Theta}\left(1 + \frac{B_i}{\Theta}\right) - \frac{B_i}{\Theta}[\Phi_{i1}\exp(\Theta\Phi_{i1}) - \Phi_{i2}\exp(\Theta\Phi_{i2})] \right\} \tag{12.5}$$

where ${}^{\infty}D_{ip}$ is the basic diffusivity of component i in the polymer and $\Theta = \ln({}^*D_{ii}/{}^{\infty}D_{ip})$.

Due to the different conditions on the retentate and permeate sides of the membrane, the local permeability varies across the thickness of the membrane. Therefore, a mean permeability $P_{f,av}$ is defined by

$$P_{f,av} = \frac{J_i}{(f_{i1} - f_{i2})} \tag{12.6}$$

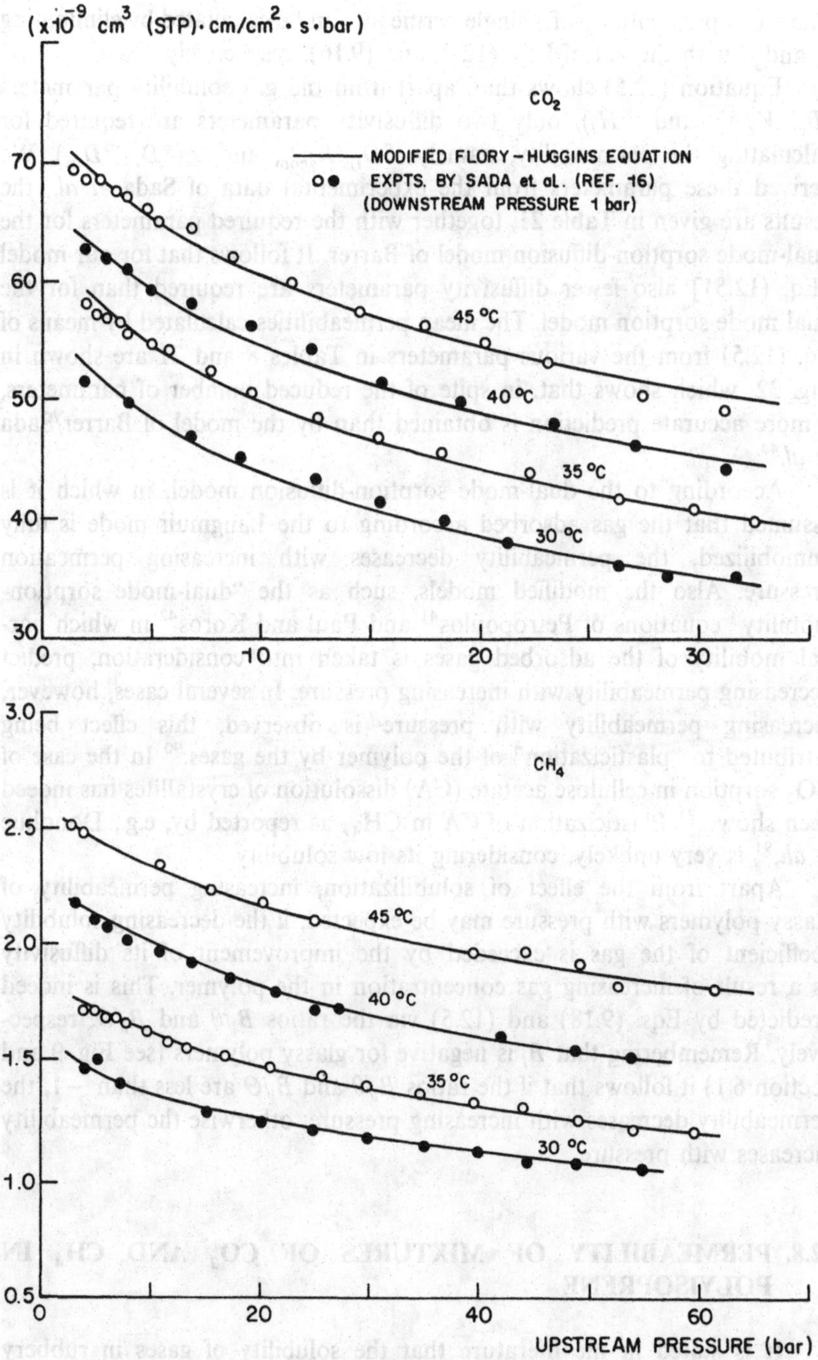

Figure 32. Prediction of gas permeability in glassy polymer.

The mean permeability of a single permeant can be calculated by eliminating J_i and f_i with the aid of Eqs. (12.5) and (9.16), respectively.

Equation (12.5) shows that, apart from the gas solubility parameters (B_i, V_i/ϕ_a, and ${}^\infty H_i$), only two diffusivity parameters are required for calculating the permeability, namely ${}^\infty D_{ip}/R_{geom}$ and $\ln({}^*D_{ii}/{}^\infty D_{ip})$. We derived these parameters from the experimental data of Sada *et al.*; the results are given in Table 21, together with the required parameters for the dual-mode sorption-diffusion model of Barrer. It follows that for our model [Eq. (12.5)] also fewer diffusivity parameters are required than for the dual-mode sorption model. The mean permeabilities calculated by means of Eq. (12.5) from the various parameters in Tables 8 and 21 are shown in Fig. 32, which shows that, in spite of the reduced number of parameters, a more accurate prediction is obtained than by the model of Barrer/Sada *et al.*[44]

According to the dual-mode sorption-diffusion model, in which it is assumed that the gas adsorbed according to the Langmuir mode is fully immobilized, the permeability decreases with increasing permeation pressure. Also the modified models, such as the "dual-mode sorption-mobility" equations of Petropoulos[41] and Paul and Koros[42] in which partial mobility of the adsorbed gases is taken into consideration, predict decreasing permeability with increasing pressure. In several cases, however, increasing permeability with pressure is observed, this effect being attributed to "plasticization" of the polymer by the gases.[90] In the case of CO_2 sorption in cellulose acetate (CA) dissolution of crystallites has indeed been shown.[21] Plasticization of CA in CH_4, as reported by, e.g., Donohue *et al.*,[91] is very unlikely, considering its low solubility.

Apart from the effect of solubilization, increasing permeability of glassy polymers with pressure may be expected, if the decreasing solubility coefficient of the gas is exceeded by the improvement of its diffusivity as a result of increasing gas concentration in the polymer. This is indeed predicted by Eqs. (9.18) and (12.5) via the ratios B_i/θ and B_i/Θ, respectively. Remembering that B_i is negative for glassy polymers (see Fig. 9 and Section 6.1) it follows that if the ratios B_i/θ and B_i/Θ are less than -1, the permeability decreases with increasing pressure; otherwise the permeability increases with pressure.

12.8. PERMEABILITY OF MIXTURES OF CO_2 AND CH_4 IN POLYISOPRENE

It is stated in the literature that the solubility of gases in rubbery polymers can be described by the linear Henry's law. This is true for relatively low volume concentrations of the gases. At larger volume fractions (say $\Phi_i \approx 0.05$) the solubility increases with pressure and, because the diffusivity increases with the concentration of the gases in a polymer as shown previously, the permeability increases also. This effect is clearly observed, e.g., with the sorption and permeation of CO_2 in polyisoprene (a rubbery polymer), showing a considerable increase in the sorption coefficient and permeability with pressure. These effects are predicted accurately by Eqs. (6.23) and (9.18) (B_i being positive in this case).

Corresponding experiments with methane in polyisoprene show almost linear sorption and permeation rate with pressure, due to the low solubility of this gas. As a consequence, the separation factor (S^i_j is the selectivity), which is defined as the permeability ratio of pure CO_2 and

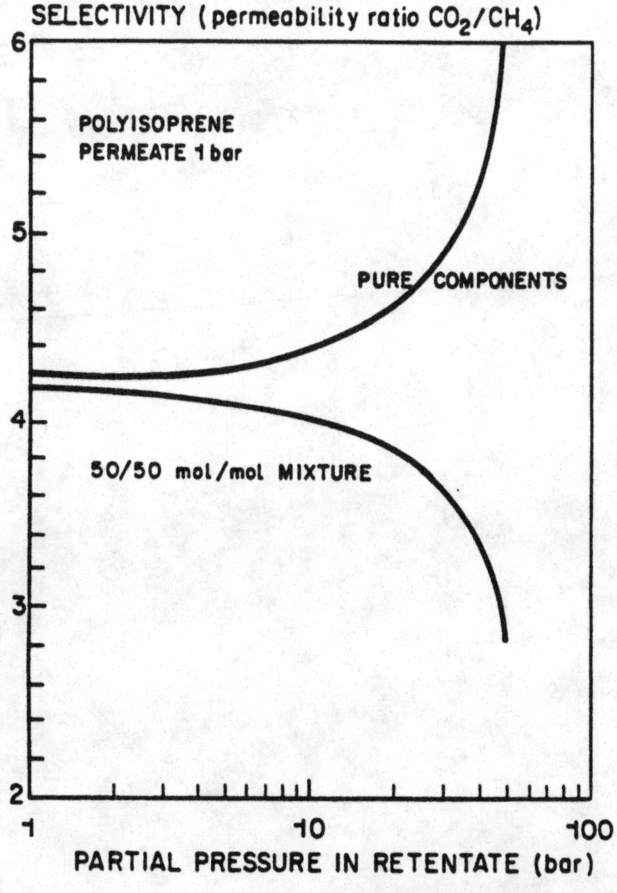

Figure 33. Effect of gas composition and pressure on selectivity of membrane separation.

CH_4, increases strongly with pressure as shown in Fig. 33. Permeation experiments with mixtures, however, show a tremendous decrease in selectivity with increasing partial pressures of the gases involved, as shown in Fig. 33. This is due to the fact that the low solubility component hardly affects the sorption of the high solubility component, while the latter causes a tremendous increase in solubility of the former because of its solvency for that component. Also, these effects, which cannot be described by the conventional equations, are predicted quantitatively by the modified Flory–Huggins equation (6.23) for gases and the gas permeability equation (9.18). This is shown by the curves in Fig. 33, which are calculated from literature data.[56]

13

OPTIMUM CHOICE OF POLYMERS FOR MEMBRANE PREPARATION

The invention of asymmetrical membranes by Loeb and Sourirajan marked the start for the technical application of membrane separation processes to aqueous solutions. Their procedure of preparing asymmetrical membranes consists of spreading out a homogeneous polymer solution onto a flat surface, evaporating some solvent from the top layer [which causes the formation of an ultrathin tight selective polymer layer ($<0.1\ \mu m$)], and then contacting it with an antisolvent. This causes demixing of the polymer solution, forming the porous support for the ultrathin tight membrane layer.

The requirements for a polymer suited for preparing asymmetrical membranes are (1) dissolution into and demixing with suitable solvents, and (2) solidifying upon evaporation of the solvent(s). Such polymers are plastomers, which melt upon heating and solidify (crystallize) upon cooling down. When developing new membranes the main emphasis has been placed on the application of these types of polymers, which all are semicrystalline. Hardly any transport takes place through the crystalline polymer phase, so the permeability of plastomers is intrinsically low, as shown in Section 9.2; hence they are less suited as membrane materials.

Optimum flux can be achieved with noncrystalline polymers such as elastomers, which in our opinion are therefore more suitable for preparing membranes. The selectivity of such materials, however, is limited if large swelling takes place in the permeants.

The modified Flory–Huggins equation expresses the fact that swelling can be controlled by the degree of crosslinking (elastic strain term) and by controlled interaction of the permeant mixture with the polymer (heat of mixing term). An example of the latter is the application of polydimethylsiloxane elastomer (pdms) for separating mixtures of mineral oil and furfural, stemming from the furfural extraction process.[92] The oil preferentially passes the membrane, while furfural is retained, because of its low interaction with (hence low solubility in) pdms. Another example is the addition of a sufficient amount of an antisolvent to the feed mixture, thus reducing the swelling of the polymer membrane to such an extent that a high selectivity is achieved at still reasonable flux. In this way we were able to develop a membrane, based on a fluorosilicone elastomer, which is capable of separating hydrocarbon solvents from mineral oil fractions at fluxes of even 2 m^3/m^2 per day at 40 bar permeation pressure, with selectivities varying between 50 and 140, depending on the concentration and molar weight of the hydrocarbons involved.[93,94]

Alternatively, a component to be retained can be converted into a compound which is insoluble in the elastomer. Such an example is the conversion of an aromatic carbonic acid into its sodium salt, thus enabling its highly selective separation from an organic solution with phenols of about the same molecular weight, using a membrane[95] of pdms.

Selective membrane separation of gases is usually more difficult than that of liquids, because their interaction with polymers is generally much smaller than liquids, in contrast to the opinion of many scientists involved in research on gas separation membranes.

Another widely accepted opinion is that glassy polymers are preeminantly suited for preparing such membranes. This might be true for the separation of permanent gases. However, in the case of gases, which are below or in the vicinity of their critical temperature, solubilization (disruption of local orientations) may occur, reducing the selectivity with time.

A prerequisite for a technically applicable membrane is its reliability over a long term (such as a couple of years). Therefore, we believe that glassy polymers do not fulfil the requirements of a suitable membrane material.

The main parameter determining the selectivity of the membrane separation of gases is their saturation pressure (or, in the case of permanent gases, their extrapolated saturation pressure). Low saturation pressures cause high solubility.

The sorption of low-solubility gases in a polymer increases strongly in the presence of other gases at relatively high concentrations; such a situation considerably decreases the selectivity of the membrane separation. This problem can be solved by the application of highly crosslinked polymers (like polymers prepared by contact polymerization and by plasma polymerization), thus reducing the solubility of all the gases involved. The specific fluxes of these materials are very low. However, extremely thin membranes can be prepared by the techniques mentioned so that sufficiently large fluxes are still achieved. In this way very selective membranes were prepared for separating various mixtures of permanent gases, and gases below or in the vicinity of their critical temperature.[96,97]

14

DISCUSSION AND CONCLUSIONS

The unreliable prediction of membrane separation by the conventional permeation equations must be attributed mainly to the applied simplifications and assumptions made during their derivation.

The various existing physical models describing membrane permeation have now been compared. It was concluded that the "solution-diffusion model" is most suited for describing membrane transport phenomena in tight (diffusion-type) membranes. According to this model each permeant dissolves in the membrane material and passes by diffusion in response to its gradient in the thermodynamic potential.

The solubility of permeants in polymer membranes is described inadequately by the Flory–Huggins equation. The main reasons are:

1. The crystallinity of the polymers involved is ignored.
2. The limited flexibility of the polymer chain parts in a polymer network is ignored.
3. The effect of permeants on the coordination number of the swollen polymer is ignored.
4. Application of the unreliable solubility parameter model for describing mixing enthalpy.

A modified Flory–Huggins equation has been derived that obviates these shortcomings. With this latter equation quantitative prediction of the equilibrium sorption of a large variety of liquid mixtures and gas mixtures in different polymers was achieved. Also, the sorption of gases in "glassy" polymers is predicted quantitatively by the modified Flory–Huggins equation without any need of the arbitrary Langmuir term, required in the conventional dual-mode sorption model. Fewer parameters are required in order to achieve at least equal accuracy.

A weak point in the prediction of solubility is still the estimation of the enthalpy contribution, for which no reliable equation is yet available. Fundamental research is required to establish a reliable theoretical equation for predicting mixing enthalpies. For the time being it is recommended that the binary interaction parameters be used, as obtained in the original van Laar equation, which can be derived, for instance, from sorption (swelling) experiments.

The Maxwell–Stefan equation is usually employed to describe diffusive transport. In this equation binary diffusion coefficients are present implicitly; this considerably reduces its practical accessibility. Furthermore, the equation predicts constant diffusivity for binary mixtures in contrast to experimental observations. This is due to the arbitrary mixing rule applied for calculating the average frictional force exerted by the mixture on diffusing molecules.

A modified Maxwell–Stefan equation (MMS equation) has been derived without applying this arbitrary mixing rule. This equation contains the coefficient of self-diffusion in the mixture of the component considered instead of the mutual binary diffusivities of the components involved. Furthermore, the diffusivity is expressed explicitly, which enhances the accessibility and facilitates practical application of the MMS equation. For steady-state conditions this equation transforms into the Darken, Prager, and Crank equation, extended for multicomponent systems.

It is shown that the coefficient of self-diffusion of a component in a mixture can be calculated from its basic self-diffusivity and its basic binary diffusivities in the other compounds involved, by applying a modified Vigne equation extended for multicomponent systems.

Nonideal mixing behavior is described by a binary constant, which can be derived from the mixture viscosity.

The basic self-diffusivities and basic binary diffusivities of the compounds are known in the literature or can be estimated from existing correlations. The basic diffusivities of permeants in polymers must generally be determined experimentally.

Based on the MMS equation a new permeation equation has been derived in which, contrary to conventional equations, diffusive transport inside the membrane is considered, depending on the driving forces exerted externally. This equation is capable of a quantitative prediction of the individual permeant fluxes for all isothermal membrane processes, such as reverse osmosis, pervaporation, dialysis, and gas permeation. This was confirmed by experiments in polar and nonpolar systems with single and binary permeant mixtures (liquids as well as gases).

A preliminary theoretical study of the effect of a permeation pressure exerted externally on a *supported* homogeneous membrane revealed that, in general, this causes no pressure gradient inside the membrane. A constant pressure exists inside the membrane that is equal to the upstream pressure.

In *unsupported* homogeneous membranes, too, a constant pressure exists inside the membrane, but its extension depends on membrane thickness and shape and on whether the permeation pressure exerted causes a compressive or a tensile stress inside the membrane material.

The pressure distribution inside heterogeneous membranes is difficult to predict and needs further study.

For liquid mixtures a simplified procedure is proposed for calculating membrane separation of multicomponent mixtures. The technique is based on the assumption of an exponential concentration gradient inside the membrane in accordance with experimental observations. The exponential form, which differs for each composition and set of conditions, can be calculated from the boundary conditions of the membrane. An exact solution of the permeation rate equation could be derived for gases.

IRREVERSIBLE PROCESSES NEAR EQUILIBRIUM

I.1. INTRODUCTION

Thermodynamics deals with systems that are in equilibrium. If transport phenomena occur in a system, driving forces must act and hence no equilibrium exists. Therefore, transport processes can in principle not be described by thermodynamics. It is expected, however, that for the statistical theories of rate processes much background information can be drawn from thermodynamics. Furthermore, it is assumed that near equilibrium and at steady-state conditions thermodynamic treatments are applicable. Steady state is achieved when the properties and other conditions of a system do not change with time while irreversible flows of heat and/or matter take place through that system. Such steady states are treated according to the rate theories of the various processes involved. Thermodynamics does not predict the absolute rate of the flow processes but yields relations between measurable quantities, thus permitting estimation of the respective rate coefficients.

The classical theory of thermodynamics for nonisothermal systems as developed, e.g., by Carnot applies to the reversible changes which occur in the working fluid, while simultaneous dissipative processes such as friction and heat transfer by conduction to the surroundings are regarded as independent and are treated separately. It is stated that if different simultaneous processes can be separated into noninteracting mechanisms and the steady irreversible flow affects only one of these processes, then an equilibrium treatment can be applied to the others.[22] For example, leakage of the solute through a membrane will not affect the osmotic pressure in a reverse osmosis experiment provided that the composition of the solutions on both sides of the membrane is kept constant by external means. The total flux through the membrane is affected by the amount of leakage, but the transport through the membrane material itself is then independent of leakage and can be treated separately.

In practice it is often difficult to identify the noninteracting processes. However, it will be shown elsewhere in this Appendix that such identification can be avoided.

I.2. THEORY OF NEAR-EQUILIBRIUM PROCESSES

An irreversible flow of matter, heat, etc., is generally accompanied by another flow. For example, in nonisothermal systems there is always some heat conductance in addition to whatever other processes exist that transfer, e.g., some matter or electricity. Identification of the independent processes may be ambiguous and/or inconvenient, and therefore a more general theory for multiflow systems is useful. These considerations led to the development of the theory of irreversible thermodynamics for near-equilibrium processes (IT). According to this theory the basic theorems of classical thermodynamics, namely, the conservation of energy and mass and the Gibbs free-energy equation, are also valid for irreversible near-equilibrium processes. Hence, the thermodynamic variables and properties are still valid when irreversible processes are proceeding at finite rates. It is assumed that each system can be subdivided into various portions in which equilibrium exists locally, so that the temperature, pressure, entropy, etc., are unambiguous within these subdivisions. If these subdivisions can be made large enough to contain a sufficient number of molecules to have macroscopic properties, and yet small enough so that the original gradients within such portions are small, then the finally measured properties may be assigned to the original system. According to Prigogine[23] this is true when transport processes occur in the range of linear-rate laws.

I.3. ENTROPY PRODUCTION IN IRREVERSIBLE FLOW PROCESSES

A basic difference between classical thermodynamics and irreversible thermodynamics is that the entropy with reversible processes in equilibrium is constant, while it increases in irreversible processes. Consequently, in the latter type of process entropy production takes place as elucidated in the following examples.

I.3a. Entropy Production in Heat Flow

Let us consider a system comprising two isothermal regions at different temperatures T_1 an T_2. The addition of an amount of heat δq to both regions would result in an entropy change of $dS_1 = \delta q/T_1$ and $dS_2 = \delta q/T_2$, respectively. If δq is the heat that flows from region 1 to region 2, then $\delta q_1 = -\delta q_2$ and the increase in irreversible entropy dS_{irr} is given by

$$dS_{\text{irr}} = \delta q(1/T_2 - 1/T_1)$$

or, in differential form,

$$dS_{\text{irr}} = \delta q \cdot d(1/T) \tag{I.1}$$

I.3b. Entropy Production in Matter Flow

Let us next consider the flow of δn moles of some compound from a region of temperature T_1 and chemical potential μ_1 to a region of temperature T_2 and chemical potential μ_2. In addition to the partial molar enthalpy (H) of the substance itself, there is usually an extra heat effect when the substance leaves the first region while the reverse effect occurs when it enters the second region. This quantity is called the (partial) molar heat of transfer and is denoted by Q^*. Or taking this into account the partial molar entropy of the substance considered can be expressed in the form

$$S = (H + Q^*)/T - \mu/T$$

However, the first law of thermodynamics requires that the sum $H + Q^*$ is identical in both regions, so the increase in entropy for the transfer of δn moles of substance equals

$$dS_{\text{irr}} = \delta n[(\mu_1/T_1 - \mu_2/T_2) + (H + Q^*)(1/T_2 - 1/T_1)]$$

or, in differential form,

$$dS_{\text{irr}} = \delta n \cdot [-d(\mu/T) + (H + Q^*)\, d(1/T)] \tag{I.2}$$

I.3c. Entropy Production by an Electric Current

A current is a flow of electrically charged particles. It contributes to the chemical potential μ_i by a quantity $z_i F_e V_e$, in which z_i, F_e, and V_e

represent the electrical charge, the Faraday constant, and the electrical potential, respectively, expressed in appropriate units. If we examine the case of an electric current in an isothermal system, the increase in entropy due to that current is expressed by

$$dS_{\text{irr}} = \delta n \cdot F_e \, dV_e/T = I\,\delta t \cdot dV_e/T \tag{I.3}$$

where I is the current and δt its duration.

I.4. RATE OF ENTROPY PRODUCTION; DISSIPATION FUNCTION

According to the examples in the preceding section, all expressions for the entropy production equal the product of an extensive quantity (δq or δn) and an intensive quantity (i.e., some kind of potential difference). The latter quantity represents a driving force for transport and is denoted by F_i for flow i.

If we consider the rate of entropy production σ ($= dS_{\text{irr}}/dt$), then the extensive quantity must be replaced by a flux, which is denoted by J_i for flux i. Hence, the rate of entropy production can in general be expressed by

$$\sigma = \left(\sum_{i=1}^{n} J_i F_i \right) > 0 \tag{I.4}$$

However, the entropy increases during a spontaneous irreversible process, so free energy is dissipated. The rate of free-energy dissipation equals $T\sigma$, which is called the dissipation function Φ and can be written as[24]

$$\Phi = T\sigma = \left(\sum_{i=1}^{n} J_i F_i \right) > 0 \tag{I.5}$$

The generalized driving force F_i in this equation differs from that in Eq. (I.4) by a factor $1/T$. In practice, it is often more convenient to employ the dissipation function instead of the rate of entropy production.

I.5. PHENOMENOLOGICAL EQUATIONS

Let us consider a system in which we have identified fully independent flow processes so that each flux is governed by its own "conjugated" force and is independent of other forces (potentials). The fluxes increase with increasing driving forces but become zero when the potential differences are zero. Therefore, such fluxes can in general be represented by a power-series expansion in the driving force resulting in

$$'J_i = 'L_{i1}\,'F_i + 'L_{i2}\,'F_i^2 + 'L_{i3}\,'F_i^3 + \cdots \tag{I.6}$$

where L_i is a rate constant and the prime denotes independent flow.

In accordance with the near-equilibrium theory, the potential differences are assumed to be so small that all but the first-power terms in the driving forces vanish, so that Eq. (I.6) reduces to the linear equation

$$'J_i = 'L_i\,'F_i \tag{I.7}$$

If fluxes (J_i) and their corresponding driving forces (F_i) are written down without regard to their mutual effects, as occur generally in actual practice, they will not be independent and cross terms may arise in the phenomenological equations. According to the near-equilibrium theory these terms are also assumed to be linear and the next relationship holds:

$$J_i = \sum_{j=1}^{n} a_{ij}\,'J_i \tag{I.8}$$

Equation (I.5) [and, of course, also Eq. (I.4)] must be valid for any set of fluxes and the corresponding potentials, so it follows that

$$\begin{aligned}\Phi &= \sum_{i=1}^{n} J_i F_i = \sum_{i=1}^{n} \left(\sum_{j=1}^{n} a_{ij}\,'J_j \right) F_i \\ &= \sum_{j=1}^{n} {'J_j} \left(\sum_{i=1}^{n} a_{ij} F_i \right) = \sum_{j=1}^{n} {'J_j}\,'F_j \end{aligned} \tag{I.9}$$

where

$$'F_j = \sum_{i=1}^{n} a_{ij} F_i \tag{I.10}$$

Combination of Eqs. (I.7), (I.8), and (I.10) yields

$$J_i = \sum_{j=1}^{n} a_{ij}\,'L_j\,'F_j = \sum_{j=1}^{n} \sum_{k=1}^{n} a_{ij}\,'L_j a_{kj} F_k$$

which, by defining

$$L_{ik} = \sum_{j=1}^{n} a_{ij}' L_j a_{kj} \tag{I.11}$$

transforms into

$$J_i = \sum_{k=1}^{n} L_{ik} F_k \tag{I.12}$$

This equation is called the *linear law of Onsager*[25] (OR).

From Eq. (I.11) we derive

$$L_{ik} = \sum_{j=1}^{n} a_{ij}' L_j a_{kj} = \sum_{j=1}^{n} a_{kj}' L_j a_{ij} = L_{ki} \tag{I.13}$$

which is called the *Onsager reciprocal relation*[25] (ORR).

I.6. REQUIREMENTS OF THE PHENOMENOLOGICAL COEFFICIENTS

Elimination of J_i from the dissipation function in Eq. (I.5) by means of Eq. (I.12) yields

$$\Phi = \sum_{i=1}^{n} \left(\sum_{k=1}^{n} L_{ik} F_k \right) F_i > 0 \tag{I.14}$$

This equation must be satisfied for each condition, hence also if there exists only a single driving force. Equation (I.14) then transforms into

$$L_{ii} F_i^2 > 0$$

from which it follows that

$$L_{ii} > 0 \tag{I.15}$$

Equation (I.14) is also valid for a system consisting of two flows and two driving forces. Therefore

$$L_{ii} F_i^2 + (L_{ik} + L_{ki}) F_i F_k + L_{kk} F_i^2 > 0$$

This must even be true if the second term on the left-hand side of this expression is negative, hence

$$L_{ii}F_i^2 - |(L_{ik} + L_{ki})\, F_iF_k| + L_{kk}F_k^2 > 0$$

The latter expression can be written as

$$[\sqrt{(L_{ii})}\, F_i - \sqrt{(L_{kk})}\, F_k]^2 + 2\sqrt{(L_{ii}L_{kk})}\, F_iF_k - |(L_{ik} + L_{ki})\, F_iF_k| > 0$$

This must also be true if the term between the square brackets equals zero. Remembering that $L_{ik} = L_{ki}$ it then follows that

$$L_{ii}L_{kk} > L_{ik}^2 \tag{I.16}$$

I.7. APPLICATION OF IT/OR/ORR IN MEMBRANES; THE KEDEM–KATCHALSKY MODEL

For isothermal transport through membranes, Eq. (I.5) transforms into

$$\Phi = \sum_{i=1}^{n} J_i \operatorname{grad}(-\mu_i)$$

For steady state and one-dimensional flow (in the z-direction) this equation can be integrated to yield

$$\int_{z=0}^{l} \Phi\, dz = \Phi_{\mathrm{m}} = \int_{z=0}^{l} \sum_{i=1}^{n} J_i \operatorname{grad}(-\mu_i)\, dz$$

Consequently

$$\Phi_{\mathrm{m}} = \sum_{i=1}^{n} J_i\, \Delta\mu_i \tag{I.17}$$

In this equation Φ_{m} is the dissipation function of the whole membrane per unit area and $\Delta\mu_i$ is the difference in the chemical potential of component i across the membrane. Thus by integration the unknown gradients in the chemical potential of the permeants inside the membrane have been replaced by the differences in the chemical potential of those compounds across the membrane, which are known or can be determined.

An aqueous solution of a single solute is now examined. In this case Eq. (I.17) becomes

$$\Phi_{\mathrm{m}} = J_{\mathrm{w}}\, \Delta\mu_{\mathrm{w}} + J_{\mathrm{s}}\, \Delta\mu_{\mathrm{s}}$$

where subscripts w and s refer to water and solute, respectively. For isothermal systems the difference in chemical potential equals[22]

$$\Delta\mu_i = RT\Delta \ln a_i + V_i \Delta P$$

When this expression is substituted into the previous equation, we obtain

$$\Phi_m = (J_w V_w + J_s V_s)\,\Delta P + RT(J_w \Delta \ln a_w + J_s \Delta \ln a_s) \qquad \text{(I.18)}$$

The activity of water (a_w) in the second term on the right-hand side of this equation can be eliminated with the aid of the Lewis equation for the osmotic pressure[22] (π_w):

$$V_w \Delta\pi_w = -RT\Delta \ln a_w$$

The activity of the solute (a_s) in the second term can be eliminated by the van't Hoff equation[22] (which applies only for small solute concentrations C_s):

$$\Delta\pi_w = RT\,\Delta C_s$$

Furthermore, at small salt concentration

$$\ln a_s = \ln C_s$$

Combination of the latter three equations and Eq. (I.18) yields

$$\Phi_m = (J_w V_w + J_s V_s)\,\Delta P + (J_s/C_{s,av} - J_w V_w)\,\Delta\pi_w \qquad \text{(I.19)}$$

where $C_{s,av} = \Delta C_s/\Delta \ln C_s$ is the logarithmic mean of the concentrations of the solute across the membrane.

The quantity $J_s/C_{s,av}$ equals about the total volume flux through the membrane, so the second term in parentheses on the right-hand side of Eq. (I.18) represents the volume flux J_D of the solute with respect to the volume flux of the solvent. The first term in parentheses represents the total volume flux of water and solute J_V through the membrane. Therefore, Eq. (I.19) can be expressed in the form

$$\Phi_m = J_V\,\Delta P + J_D\,\Delta\pi_w > 0$$

Consequently, the fluxes and corresponding driving forces are known, and therefore we can write according to the OR

$$J_V = L_V \Delta P + L_{VD} \Delta \pi_w \tag{I.20}$$

and

$$J_D = L_{DV} \Delta P + L_D \Delta \pi_w \tag{I.21}$$

where, according to the ORR,

$$L_{VD} = L_{DV}$$

By defining $L_{VD}/L_V = -\sigma$, equal to the reflection (couplings) coefficient, Eq. (I.20) can be transformed into

$$J_V = L_V(\Delta P - \sigma \Delta \pi_w) \tag{I.22}$$

This first equation of Kedem and Katchalsky[26] predicts the magnitude of the total volume flux through the membrane; L_V is called the filtration coefficient. The reflection coefficient indicates the solute rejection properties of the membrane. If σ is zero, the membrane is completely permeable to solute and solvent; if $\sigma = 1$, the membrane is completely impermeable to the solute.

An equation for the solute flux can be derived as follows. By summing J_D and J_V using Eq. (I.19) and assuming that

$$V_s C_{s,av} \ll 1$$

it is found that

$$J_s = C_{s,av}(J_V + J_D)$$

Elimination of J_D by means of Eq. (I.21) and ΔP by means of Eq. (I.22) leads to

$$J_s = C_{s,av}(1 - \sigma) J_V + C_{s,av} \omega \Delta \pi_w \tag{I.23}$$

where $\omega = (L_V L_D - L_{VD}^2)/L_V$ is the solute permeability at zero volume flux.

Equation (I.23) is the second equation of Kedem and Katchalsky.[26] This equation predicts that the solute flux is proportional to the logarithmic mean of the solute concentration across the membrane and to the total volume flux through it.

II

DERIVATION OF THE CONCENTRATION POLARIZATION EQUATION OF GASES

II.1. CONTINUITY EQUATION

The general equation of continuity for substance i reads[98]

$$\nabla\rho_i \mathrm{U} - M_i \nabla^2(D_{im} C_i) + \frac{\partial \rho_i}{\partial t} = M_i Q_i \tag{II.1}$$

where U is the linear velocity [L/t], ρ the density [M/L^3], C the concentration [mol/L^3], D_{im} the diffusivity [L^2/t], M_i the mole weight of i [M/mol], and Q_i is the production rate of i [mol/L$_3$/t].

Now

$$C_i = \rho_i / M_i \tag{II.2}$$

If Eq. (II.2) is substituted into Eq. (II.1) and we assume $Q_i = 0$ and steady-state conditions, the latter equation transforms, for one-dimensional flow (in the z-direction), into

$$\frac{\partial C_i U_z}{\partial z} - \frac{\partial^2 D_{im} C_i}{\partial z^2} = 0 \tag{II.3}$$

II.2. TRANSPORT EQUATION FOR TIGHT MEMBRANES

Figure 34 shows schematically the various gas flows to and from a tight membrane, where substance i is being retained. We first consider the

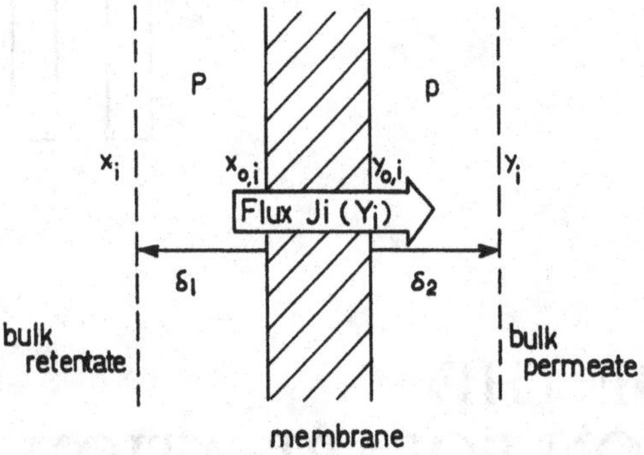

Figure 34. Model for membrane separation of gases.

flows at the retentate side and assume that there is no pressure drop from the bulk phase to the membrane; hence P is constant. Assuming ideal gas behavior we may set

$$U = J\frac{P_0}{P} \tag{II.4}$$

$$C_i = P_i/RT \tag{II.5}$$

and

$$x_i = P_i/P \tag{II.6}$$

where x_i is the mole fraction of i in the retentate, P_0 the pressure at standard conditions (e.g., 1 bar), P_i the partial pressure of i in the retentate, and J is the total flux in $m^3(STP)/(m^2 \cdot day)$.

According to kinetic gas theory the diffusivity D_{im} is inversely proportional to the total pressure P and depends on the constituents of the gas mixture but hardly changes with their concentrations.[99] Therefore we define

$$D_0 = D_{im}P/P_0 = \text{constant} \qquad [L^2/t] \tag{II.7}$$

where D_0 equals the diffusivity at one bar.

Substitution of Eqs. (II.4)–(II.7) into Eq. (II.3) yields, after manipulation,

$$-D_0\frac{\partial^2 x_i}{\partial z^2} + J\frac{\partial x_i}{\partial z} = 0 \tag{II.8}$$

Integration yields

$$-D_0 \frac{dx_i}{dz} + Jx_i = J_i = \text{constant} \tag{II.9}$$

where J_i is the flux of substance i through the membrane, in m^3(STP)/($m^2 \cdot$ day).

The solution of this differential equation is

$$x_i = J_i/J + k \exp\left(\frac{J}{D_0} z\right) \tag{II.10}$$

where k is the integration constant, which can be determined from the boundary conditions

$$x_i = x_i, \qquad z = -\delta,$$

$$x_i = x_{0,i}, \qquad z = 0,$$

yielding

$$k = x_{0,i} - Y_i \tag{II.11}$$

with δ the thickness of the laminar gas layer (stagnant zone) along the membrane surface, $x_{0,i}$ the concentration of substance i at the retentate side boundary of the membrane (mole fraction), and Y_i the local permeate concentration of i (mole fraction).

On combining Eqs. (II.10) and (II.11) we obtain

$$x_i = Y_i + (x_{0,i} - Y_i) \exp\left(-\frac{J}{D_0} \delta_1\right) \tag{II.12}$$

where 1 denotes retentate side.

We now consider the various flows at the permeate side. It can be derived from Fig. 34 that Eq. (II.9) transforms into

$$-D_0 \frac{dy_i}{dz} + Jy_i = J_i = JY_i = \text{constant} \tag{II.13}$$

if y_i is the concentration of i in the bulk permeate phase.

The approach described above enables this differential equation to be solved. After integration between the boundaries

$$y_i = y_{0,i}, \qquad z = 0,$$

$$y_i = y_i, \qquad z = \delta_2,$$

the equation

$$y_i = Y_i + (y_{0,i} - Y_i)\exp\left(\frac{J}{D_0}\delta_2\right) \tag{II.14}$$

is obtained, where $y_{0,i}$ is the concentration of substance i at the permeate-side boundary of the membrane (mole fraction) and 2 denotes the permeate side.

According to Eq. (3.22) in Chapter 3 the local flux of i through the membrane is expressed by

$$J_i = K_i(Px_{0,i} - py_{0,i}) \tag{II.15}$$

With the aid of Eqs. (II.12) and (II.14) quantities $x_{0,i}$ and $y_{0,i}$ in Eq. (II.15) can be eliminated to finally yield

$$x_i = Y_i\left[1 + \left(\frac{J}{K_i(P-p)} - 1\right)\left(1 - \frac{p}{P}\right)\exp\left(-\frac{J}{b_1}\right)\right] - \frac{p}{P}(Y_i - y_i)\exp\left(-\frac{J}{b_1} - \frac{J}{b_2}\right) \tag{II.16}$$

where $b_1 = D_0/\delta_1$ is the backdiffusion factor at the retentate side of the membrane and $b_2 = D_0/\delta_2$ is the backdiffusion factor at the permeate side of the membrane.

II.3. EXPERIMENTS AND RESULTS

The magnitude of the backdiffusion factor (b) was estimated by performing permeation experiments with mixtures of H_2 and CO_2 (ranging from 25 to 75%v of H_2), applying porous polypropylene membranes (Celgard 2402, five-ply layers). Such membranes exhibit Knudsen selectivity, implying that the flow rate is inversely proportional to the square root of the molar mass of the gases involved.

The membranes were clamped between two flanges into which spiral-shaped grooves of mirror-image form had been cut, such that the channels formed on either side of the membrane permitted countercurrent flow of the retentate and permeate.

The cross-section of the channels equalled $4 \times 4\,\text{mm}^2$. However, in order to support the membrane a wire with a cross-sectional diameter of 4 mm was put into the groove on the permeate side, which reduced the cross-sectional area to about 20%. As a result, the flow rate along the membrane at the permeate side was so large for these experiments that the exponential function in the second term on the right-hand side of Eq. (II.16) appeared to approach unity.

Experiments were performed with groove lengths varying between 40 and 165 cm. The upstream pressure was varied between 2 and 11 bar, while the downstream pressure was kept atmospheric. The flow rates and compositions of feed, retentate, and permeate were measured, from which the backdiffusion factors were calculated by means of Eq. (II.16) using an iteration procedure.

The results are collected in Fig. 35, where the ratio of the backdiffusion factor to diffusivity at atmospheric pressure is plotted versus the average normal volume flow rate (q_{av}) of the gas mixture along the mem-

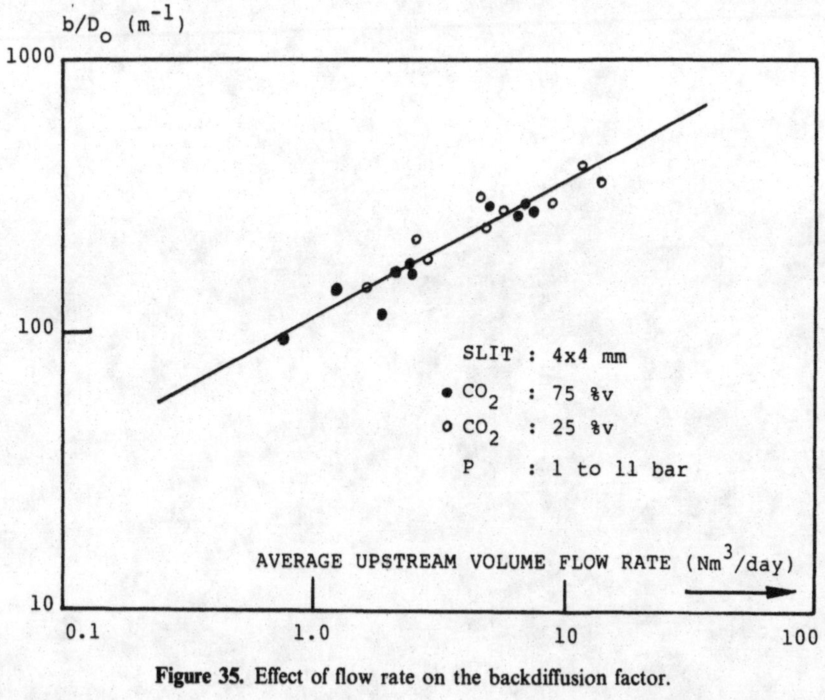

Figure 35. Effect of flow rate on the backdiffusion factor.

brane surface. It is seen that within the experimental accuracy all the ratios coincide with the same curve, which indicates a square-root dependence on q_{av}, independent on groove length, total pressure, and gas composition.

It can be derived from these data and the cross-sectional area of the groove that the backdiffusion factor can be expressed by the empirical relation

$$b = 136 D_0 (PU)^{1/2} \quad [\mathrm{m/s}] \tag{II.17}$$

where U is the linear gas velocity at pressure P [m/s], P the absolute pressure [bar], and $D_0 = D_{im} P/P_0$ [m^2/s].

Quantity b is independent of gas-mixture composition and, furthermore, U is inversely proportional to the absolute pressure. It therefore follows that b is proportional to the molecular flow rate (mol/s) along the membrane.

III

DERIVATION OF THE MAXWELL–STEFAN EQUATION

A molecule i in a multicomponent mixture moves in that mixture under the influence of a driving force which can, in general, be represented by a gradient in the thermodynamic potential grad μ_i. As a result the mixture molecules exert on that molecule a frictional force f, which opposes the driving force and becomes equal to it after molecule i has reached its steady-state velocity. Then

$$f = -\text{grad}\,\mu_i \tag{III.1}$$

It is assumed that the frictional force exerted by a molecule j on molecule i is proportional to their difference in velocity w, hence

$$f_{ij} = \frac{RT}{\boldsymbol{D}_{ij}}(w_i - w_j) \tag{III.2}$$

if $RT/\boldsymbol{D}_{ij}$ represents a friction coefficient.

Let us consider a system of s different components and assume that the molecule i collides during its movement through that system with n_j molecules j, n_k molecules k,..., n_s molecules s. Then the average force exerted on molecule i is given by

$$f_{i,\text{av}} = \sum_{j=1}^{s}\left[\sum_{k=1}^{n_j}\frac{RT}{\boldsymbol{D}_{ij(k)}}(w_i - w_j)_k\right]\Big/\sum_{j=1}^{s} n_j \tag{III.3}$$

It must be emphasized that w_i may change after each collision and also that w_j as well as $\boldsymbol{D}_{ij(k)}$ may be different for each collision.

We now make the following assumptions: The friction coefficient $RT/D_{ij(k)}$ is independent of the presence of other components and is constant for each collision (k) between molecules i and j.

It is assumed that so many collisions are considered that

$$n_j \Big/ \sum_{j=1}^{s} n_j = x_j = \text{local concentration of } j \tag{III.4}$$

and furthermore that

$$\sum_{k=1}^{n_j} (w_i - w_j)_k \Big/ \sum_{j=1}^{s} n_j = x_j(w_{i,\text{av}} - w_{j,\text{av}}) \tag{III.5}$$

where $w_{i,\text{av}}$ and $w_{j,\text{av}}$ represent the average velocities of molecules i and j in the mixture, respectively.

Combination of the various equations then yields

$$f_{i,\text{av}} = \sum_{j=1}^{s} \frac{RT}{D_{ij}} x_j(w_{i,\text{av}} - w_{j,\text{av}}) = -\text{grad}\,\mu_i \tag{III.6}$$

If the total flow of molecules i and j is N_i and N_j, respectively, then

$$N_i = x_i C w_{i,\text{av}} = C_i w_{i,\text{av}} \text{ etc.}$$

where C is the total concentration and C_i the concentration of molecules i.

Substitution in Eq. (III.6) yields the Maxwell–Stefan equation:

$$\frac{\text{grad}\,\mu_i}{RT} C_i = \sum_{\substack{j=1 \\ j\neq i}}^{s} \frac{x_i N_j - x_j N_i}{D_{ij}} \tag{III.7}$$

IV

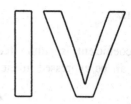

DERIVATION OF THE MODIFIED MAXWELL–STEFAN EQUATION

In the derivation of the Maxwell–Stefan equation it was assumed that the interaction between molecules i and j is unaffected by the presence of other compounds, hence the mutual friction coefficient $RT/\boldsymbol{D}_{ij}$ is independent of concentration and composition. This implies that at the moment of a collision between i and j no other molecules are involved. This may be valid in gas mixtures but it is certainly not true in liquid mixtures, as is expressed clearly in Fig. 7.

This figure shows that, in the case of gases, usually only two molecules are involved per collision. Therefore, one may expect that the average friction coefficient in a gas mixture can indeed be described by the molar average of the binary friction coefficients of i and the other participating molecules. It is also shown in Fig. 7 that the situation in liquid mixtures is quite different, because then each molecule is always surrounded by a relatively large number of "direct neighbor molecules," which is expressed by the coordination number Z_{av}. Therefore, a molecule i in a liquid mixture will always experience a frictional force resulting from the *combined interactions* of all direct surrounding molecules, which for the *same local composition* is described by the *same single mixture friction coefficient* and which we define as $\sigma_{i\mathrm{m}}\eta_{\mathrm{m}}$. As a result, Eq. (III.3) in Appendix III then transforms into

$$f_{i\mathrm{m}} = \sigma_{i\mathrm{m}}\eta_{\mathrm{m}} \sum_{j=1}^{s} \left[\sum_{k=1}^{n_j} (w_i - w_j)_k \right] \Big/ \sum_{j=1}^{s} n_j \qquad \text{(IV.1)}$$

If Eqs. (III.4) and (III.5) are assumed to be valid and the friction coefficient is eliminated by $RT/{}^*D_{im}$ [see Eq. (8.11)], then Eq. (IV.1) can be expressed in the form

$$C_i \frac{{}^*D_{im}}{RT} \operatorname{grad} \mu_i = \sum_{\substack{j=1 \\ j \neq i}}^{s} (x_i N_j - x_j N_i) \tag{IV.2}$$

which, apart from the coefficient of self-diffusion in the mixture, resembles the Maxwell–Stefan equation.

V

THEORY OF THE ENTROPY OF MIXING

V.1. DERIVATION OF THE ENTROPY EQUATION

V.1a. Procedure

In order to calculate the configuration entropy of a swollen polymer system we apply Huggin's procedure,[50] namely, we count the number of possible arrangements of the molecules involved at the sites of a quasi-lattice. If the number of distinguishable configurations of the mixture of molecules is denoted by Ω and that of the individual components by $\Omega_{0,j}$, then the entropy of mixing Δs^{M} is found by using the well-known Boltzmann equation

$$\Delta s^{\mathrm{M}} = R\left(\ln \Omega - \sum_j \ln \Omega_{0,j}\right) \qquad \text{(V.1)}$$

A system of polymer molecules, which are partly crystallized and partly in the amorphous state, is examined and it is assumed that no crystals dissolve during swelling. The permeating molecules are present in the amorphous polymer phase while the crystallites form a separate phase (see Fig. 36), so only the first phase needs to be considered. The problem is then to count the number of possible configurations of the segments of the amorphous chains, which all start on the surface of certain crystals and end at the surface of certain crystals, without defining the position of these crystals. The problem is solved by successively placing the amorphous polymer chains in a lattice consisting of V_t sites each of volume equal to the polymer segment (which can be chosen arbitrarily) and counting the

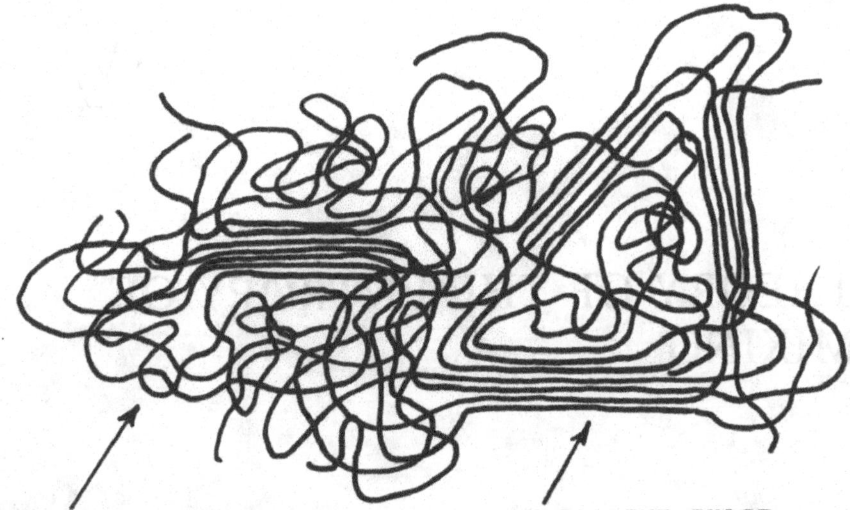

Figure 36. Model of semicrystalline polymers.

number of possible placings of each segment. The same procedure is applied for the penetrants, which are assumed to consist of segments of identical size.

V.1b. Configurations of Polymer Chains

We count the number of possible configurations of the qth amorphous polymer chain after placing $(q-1)$ chains. Owing to the polymer structure, the strain (due to the fact that the chains must start and end on crystals), etc., the number of sites which a segment can occupy with respect to its direct predecessor is restricted to, let us say, Y, which we call flexibility (see Fig. 37). This number Y must further be corrected for the chance that some of the sites have been occupied already by previously placed segments from either the same or previously placed chains.

According to Huggins[50] this correction $F(q_p)$ is given by

$$F(q_p) = [1 - f(q)] \Big/ \left[1 - \frac{2}{Z_{av}} f(q)\right] \tag{V.2}$$

where

$$f(q) = \frac{(q-1)\, L_{av}}{V_t}\,[1 - f(0)] + f(0) \tag{V.3}$$

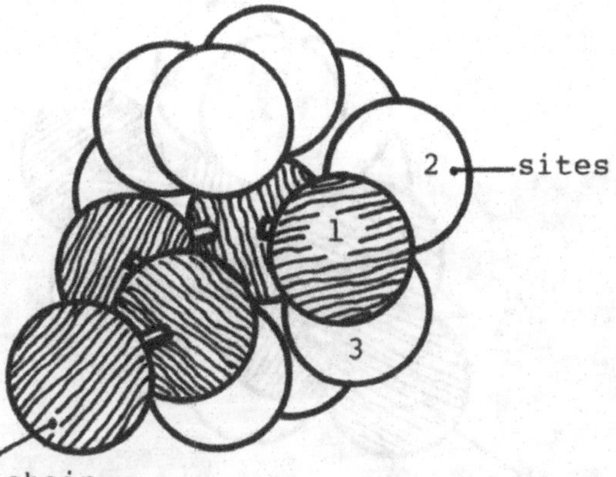

Figure 37. Flexibility of polymer segments. Segment 1 of the polymer chain can also occupy site 2 or 3, hence $Y=3$.

Z_{av} is the coordination number of the swollen polymer phase, L_{av} the average number of segments per amorphous polymer chain, and $f(0)$ the chance that a site has been occupied by a segment of the chain being placed. Huggins holds that $f(0)$ is negligible for chains smaller than about eight segments and approaches a small constant (≈ 0.02) for larger chains.

The numerator of Eq. (V.2) represents the fraction of unoccupied lattice sites, while the denominator is a correction for the fact that, if a certain site has been occupied by a segment of a polymer chain, at least two of the surrounding Z_{av} sites must also be occupied by segments from the same chain (see Fig. 38). In general, the number of arrangements of the nth segment of the qth amorphous polymer chain (Ω_{qn}) is expressed by

$$\Omega_{qn} = YF(q_p) \tag{V.4}$$

If the number of segments of the qth amorphous polymer chain is L, then the total number of arrangements of all segments of the qth chain (Ω_q) is

$$\Omega_q = \prod_L \Omega_{qn} = Y^L F(q_p)^L \tag{V.5}$$

If the number of amorphous polymer chains in the system considered equals N_p and the number of segments per chain follows a distribution function $W(L)\,dL$, such that

$$\int_{L_{min}}^{L_{max}} W(L)\,dL = 1 \tag{V.6}$$

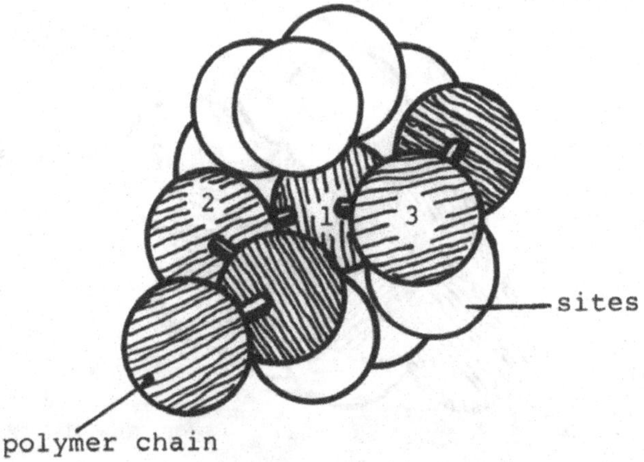

Figure 38. Quasi-lattice of a polymer system. If site 1 is occupied by an intermediate segment of a polymer chain (segment 1), at least two of the Z surrounding sites are occupied by segments of the same chain (namely, segments 2 and 3).

then the number of amorphous polymer chains of length L equals $N_p W(L)\,dL$ and the total number of arrangements of all chains of length L (Ω_L) is then

$$\Omega_L = \prod_t \Omega_t = Y^{LN_p W(L)\,dL} \prod_t [F(t_p)^L] \tag{V.7}$$

In this equation t can adopt every value between zero and N_p. As all chains of different length (number of segments) are placed randomly in the lattice, the product

$$\prod [F(t_p)^L]_{t=q,r,s\text{ etc.}}$$

must be independent of the sequence in which the chains of different lengths are placed; hence, the average value of $F(t_p)$ is the same for all chain lengths. This is only the case if

$$\prod_t F(t_p) = \prod_1^{N_p W(L)\,dL} \left[\prod_{q=1}^{N_p} F(q_p)\right]^{1/N_p} \tag{V.8}$$

As a consequence, the total number of arrangements of all amorphous polymer chains (Ω_p) is given by

$$\Omega_p = \prod \Omega_L = Y^{L_{av} N_p} \prod_{q=1}^{N_p} [F(q_p)^{L_{av}}] \tag{V.9}$$

V.1c. Configurations of Penetrants

We assume that there are N_i penetrating molecules of n_i segments, N_j penetrating molecules of n_j segments, etc., in the system under consideration. After the amorphous polymer chains are placed in the lattice, there are still N_s sites free in the lattice for the penetrating molecules.

After (q_s-1) such molecules are placed, the number of arrangements of the first segment of the q_sth penetrating molecule Ωq_{s1} is the same as the number of unoccupied sites:

$$\Omega q_{s1} = n_{\text{av}} N_s \left[1 - \frac{(q-1)}{N_s}\right] \tag{V.10}$$

where

$$n_{\text{av}} N_s = \sum_i n_i N_i \tag{V.11}$$

if n_{av} is the average number of segments per penetrating molecule and N_s the total number of penetrating molecules.

The second segment of this molecule can occupy one of the Z_{av} surrounding sites if they are not occupied by segments of previously placed molecules.

We adopt Huggins's reasoning again[50] in order to calculate the chance of occupation. The chance that a site S_0 is occupied by a segment of an amorphous polymer chain is $N_{\text{p}} L_{\text{av}}/V_{\text{t}}$. However, at least two adjacent sites must then be occupied too (see Fig. 38), so the chance that a certain adjacent site has been occupied by a segment of the same polymer chain is

$$\frac{2}{Z_{\text{av}}} \cdot \frac{L_{\text{av}} N_{\text{p}}}{V_{\text{t}}}$$

If S_0 is occupied by a segment of a penetrating molecule, at least one of the Z_{av} surrounding sites is occupied if that segment is an "end" segment (see Fig. 39), or two such sites if it is an intermediate one. As there are normally two "end" segments and $(n_{\text{av}}-2)$ intermediate ones in such molecules, the chance that a certain adjacent site of S_0 is occupied by the same penetrating molecule is

$$n_{\text{av}} \left[\frac{2}{n_{\text{av}}} \frac{1}{Z_{\text{av}}} + \frac{(n_{\text{av}}-2)}{n_{\text{av}}} \frac{2}{Z_{\text{av}}}\right] \frac{(q_s-1)}{V_{\text{t}}} = (n_{\text{av}}-1) \frac{2}{Z_{\text{av}}} \frac{(q_s-1)}{V_{\text{t}}}$$

Consequently, the chance that a certain adjacent site of S_0 is occupied by

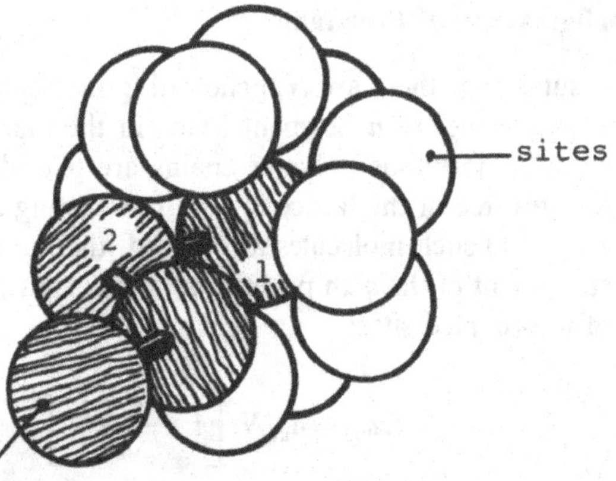

Figure 39. Quasi-lattice of a polymer system. If a site (e.g., 1) is occupied by an end segment of a chain (segment 1), at least one of the Z surrounding sites ($Z = 12$) is also occupied by a segment of the same chain (namely, segment 2).

a segment of a previously placed polymer chain or penetrating molecule which also occupies S_0 is

$$\frac{2}{Z_{\text{av}}}\left[\frac{N_{\text{p}}L_{\text{av}} + (n_{\text{av}} - 1)(q_s - 1)}{V_{\text{t}}}\right]$$

and the chance that the adjacent site is not occupied if S_0 is occupied is

$$1 - \frac{2}{Z_{\text{av}}}\left[\frac{N_{\text{p}}L_{\text{av}} + (n_{\text{av}} - 1)(q_s - 1)}{V_{\text{t}}}\right]$$

The chance that S_0 is not occupied is the same as the fraction of unoccupied sites, namely,

$$1 - \frac{N_{\text{p}}L_{\text{av}} + n_{\text{av}}(q_s - 1)}{V_{\text{t}}}$$

It will be clear that the chance $F(q_s)$ that an adjacent site of S_0 is free if S_0 is free as well is larger than if S_0 is occupied and is, as Huggins[50] pointed out,

$$F(q_s) = \frac{\left[1 - \dfrac{N_{\text{p}}L_{\text{av}} + n_{\text{av}}(q_s - 1)}{V_{\text{t}}}\right]}{\left\{1 - \dfrac{2}{Z_{\text{av}}}\left[\dfrac{N_{\text{p}}L_{\text{av}} + (n_{\text{av}} - 1)(q_s - 1)}{V_{\text{t}}}\right]\right\}} \tag{V.12}$$

Consequently, the number of arrangements of the second segment of the penetrating molecule Ωq_{s2} is given by

$$\Omega q_{s2} = Z_{\text{av}} F(q_s) \tag{V.13}$$

The number of arrangements of the other segments in that molecule, Ωq_{sn}, is restricted by its flexibility (y_i) and the chance of being occupied, hence

$$\Omega q_{sn} = y_i F(q_s)$$

Hence the number of configurations of the q_sth molecule consisting of n_i segments is

$$\Omega q_s = Z_{\text{av}}\, y_i^{(n_i-2)} n_{\text{av}} N_s \left(1 - \frac{q_s - 1}{N_s}\right) F(q_s)^{(n_i-1)} \tag{V.14}$$

and the total number of configurations of all molecules of type i is then given by

$$\Omega_i = \prod_{N_i} \Omega q_s = Z_{\text{av}}^{N_i}\, y_i^{(n_i-2)\,N_i} (n_{\text{av}} N_s)^{N_i} \times \prod_t \left(1 - \frac{t-1}{N_s}\right) \prod_t F(t_s)^{(n_i-1)} \tag{V.15}$$

where t can adopt any value between zero and N_s.

For the same reasons as argued for amorphous polymer chains [see Eq. (V.8)], the average value of $F(t_s)$ must be the same for all types of penetrating molecules present. This is true only if

$$\prod_t F(t_s) = \left[\prod_{q_s=1}^{N_s} F(q_s)\right]^{N_i/N_s} \tag{V.16}$$

Therefore, the total number of configuration of all penetrating molecules Ω_s is

$$\Omega_s = \prod_i \Omega_i = Z_{\text{av}}^{N_s} \prod_i \left[y_i^{(n_i-2)\,N_i}\right] (n_{\text{av}} N_s)^{N_s} \times \prod_{q_s=1}^{N_s} \left[\left(1 - \frac{q_s - 1}{N_s}\right) F(q_s)^{(n_{\text{av}}-1)}\right] \tag{V.17}$$

V.1d. Configurations Before Mixing

For an unswollen polymer, the coordination number and flexibility will usually differ from those of a swollen polymer. If these quantities are denoted by Z_a and y, respectively, the number of configurations of the amorphous polymer chains before swelling, Ω_{p0}, is found by substituting in Eq. (V.9): Z_a for Z_{av}, y for Y, $L_{av}N_p$ for V_t, $f(q_0)$ for $f(q)$, and $f(q_{p0})$ for $f(q_p)$, where

$$f(q_0)=\frac{q-1}{N_p}[1-f(0)]+f(0) \tag{V.18}$$

and

$$f(q_{p0})=[1-f(q_0)]\bigg/\left[1-\frac{2}{Z_a}f(q_0)\right] \tag{V.19}$$

Hence

$$\Omega_{p0}=y^{L_{av}N_p}\prod_{q=1}^{N_p}F(q_{p0})^{L_{av}} \tag{V.20}$$

It will be clear that the coordination numbers of the pure penetrants i, j, etc., will also usually deviate from that of the swollen polymer. We denote them by Z_i, Z_j, etc., respectively. The number of configurations of each of the separate pure components (Ω_{i0}, Ω_{j0}, etc.) is found by substituting in Eq. (V.17): $L_{av}N_p=0$, n_iN_i for V_t, N_i for N_s, n_i for n_{av}, and Z_i for Z_{av}, resulting in

$$\prod_i \Omega_{i0}=\prod_i\left\{Z_i^{N_i}y_i^{(n_i-2)N_i}(n_iN_i)^{N_i}\right.$$
$$\left.\times\prod_{q_s=1}^{N_i}\left[\left(1-\frac{q_s-1}{N_i}\right)F(q_i)^{(n_i-1)}\right]\right\} \tag{V.21}$$

where

$$F(q_i)=\left(1-\frac{q_s-1}{N_i}\right)\bigg/\left[1-\frac{2}{Z_i}\left(\frac{n_i-1}{n_i}\right)\left(\frac{q_s-1}{N_i}\right)\right]$$

V.1e. Entropy of Mixing

The total number of distinguishable configurations of the swollen polymer system (Ω) and the separate components (Ω_0) is given, respectively, by

$$\Omega=\frac{\Omega_s\Omega_p}{(N_p!)\prod_i(N_i!)} \tag{V.22}$$

and

$$\Omega_0 = \frac{\Omega_{p0} \prod_i \Omega_{i0}}{(N_p!) \prod_i (N_i!)} \tag{V.23}$$

where the terms in the denominator of these equations eliminate counting distinguishable configurations. Then, according to Boltzmann, the entropy of mixing Δs^M equals

$$\Delta s^M = R(\ln \Omega - \ln \Omega_0) = R\left(\ln \Omega_p + \ln \Omega_s - \ln \Omega_{p0} - \sum_i \ln \Omega_{i0} \right) \tag{V.24}$$

Elimination of Ω_p, Ω_s, Ω_{p0}, and Ω_{i0} in Eq. (V.24) by means of Eqs. (V.9), (V.17), (V.20), and (V.21), respectively, applying Stirling's approximation, and setting

$$\sum_{q=1}^{N} \ln[1 - a(q-1)] = \int_{q=0}^{N} (1 - aq)\, dq \tag{V.25}$$

results in the equation

$$\begin{aligned}
\frac{\Delta s_M}{R} = {} & L_{av} N_p \ln \frac{Y}{y} + N_s \ln Z_{av} - \sum_i N_i \ln Z_i + N_s \ln(n_{av} N_s) - \sum_i N_i \ln(n_i N_i) \\
& - N_s \ln\left(\frac{n_{av} N_s}{V_t}\right) + L_{av} N_p \ln\left\{\left[1 - \frac{2}{Z_a} f(0)\right] \Big/ \left[1 - \frac{2}{Z_{av}} f(0)\right]\right\} \\
& + \left[\left(\frac{Z_{av}}{2} - 1\right) V_t + N_s\right] \ln\left[1 - \frac{2}{Z_{av}}\left(1 - \frac{N_s}{V_t}\right)\right] \\
& - \frac{L_{av} N_p}{b_0} (1 - b_0) \ln(1 - b_0) \\
& - \sum_i \left[\left(\frac{Z_i}{2} - 1\right) n_i N_i + N_i\right] \ln\left[1 - \frac{2}{Z_i} \frac{(n_i - 1)}{n_i}\right] \\
& + \left(\frac{V_t}{b} - L_{av} N_p\right) \ln\left(1 - b \frac{L_{av} N_p}{V_t}\right) \\
& - \left(\frac{Z_{av}}{2} V_t - L_{av} N_p\right) \ln\left(1 - \frac{2}{Z_{av}} \frac{L_{av} N_p}{V_t}\right)
\end{aligned} \tag{V.26}$$

where

$$b = \frac{2}{Z_{av}} [1 - f(0)] \Big/ \left[1 - \frac{2}{Z_{av}} f(0)\right] \approx \frac{2}{Z_{av}}$$

and

$$b_0 = \frac{2}{Z_a}[1 - f(0)] \Bigg/ \left[1 - \frac{2}{Z_a} f(0)\right] \approx \frac{2}{Z_a}$$

As $f(0)$ is very small, b is almost $2/Z_{av}$. Therefore, the last two terms of Eq. (V.26) cancel each other out.

V.1f. Partial Molar Entropy of Mixing

The partial molar entropy change of component i in the system (ΔS_i^{Mix}) is found by partial differentiation of Eq. (V.26) with respect to N_i, keeping the amounts of polymer and other compounds constant. The parameters $Z_a, Z_i, Z_j, \ldots, L_{av}, n_i, n_j, \ldots, y$ are independent of N_i but Z_{av}, n_{av}, and Y are not. Hence in order to perform the differentiation, the dependence of the last three quantities on N_i must be known.

The average number of segments per penetrating molecule is given by

$$n_{av} = \sum_i n_i N_i / N_s \tag{V.27}$$

where

$$N_s = \sum_i N_i$$

We now assume that Z_{av} is the volume average of the coordination numbers of all participating molecules before mixing, hence

$$Z_{av} = \sum_j n_j N_j Z_j / V_t \tag{V.28}$$

where

$$\sum_j n_j N_j = (L_{av} N_p + n_{av} N_s) = V_t$$

The dependence of Y on N_i will be considered separately in Section V.2 of this Appendix.

We define

$$\Phi_i = \frac{n_i N_i}{V_t} = \text{volume fraction of component } i \text{ in the amorphous polymer phase}$$

and

$$\Phi_s = \sum_i \left(\frac{n_i N_i}{V_t}\right) = \text{volume fraction of penetrants in the amorphous polymer phase}$$

Differentiation of Eq. (V.26) and subsequent manipulation then lead to

$$\frac{\Delta S_i^{\text{Mix}}}{R} = -\ln\left(\frac{Z_i}{Z_{\text{av}}}\Phi_i\right) - n_i\left(1 - \frac{Z_i}{Z_{\text{av}}}\right) + L_{\text{av}} N_{\text{p}} \times \left\{d \ln \frac{Y}{y} \Big/ dN_i\right\}_{N_{\text{p}}, N_j \cdots} + \left[\left(\frac{Z_i}{2} - 1\right) n_i + 1\right] \times \ln\left\{1 + \frac{n_i\left(1 - \frac{Z_i}{Z_{\text{av}}}\right) - \left(1 - \frac{Z_i}{Z_{\text{av}}}\frac{n_i}{n_{\text{av}}}\Phi_s\right)}{\left[\left(\frac{Z_i}{2} - 1\right) n_i + 1\right]}\right\} \tag{V.29}$$

As the fraction under the logarithmic sign of the last term in this equation is generally much smaller than unity, we may expand the logarithm and ignore squared and higher terms. Then Eq. (V.29) becomes

$$\frac{\Delta S_i^{\text{Mix}}}{R} = -\ln\left(\frac{Z_i}{Z_{\text{av}}}\Phi_i\right) - 1 + \frac{Z_i}{Z_{\text{av}}}\frac{n_i}{n_{\text{av}}}\Phi_s + L_{\text{av}} N_{\text{p}} \left\{d \ln \frac{Y}{y} \Big/ dN_i\right\}_{N_{\text{p}}, N_j \cdots} \tag{V.30}$$

V.2. ENTROPY OF ELASTIC STRAIN

The first term of the right-hand side of Eq. (V.26) represents in fact a contribution to the entropy of mixing due to elastic stretching of the amorphous polymer chains by swelling (Δs^{EL}):

$$\Delta s^{\text{EL}} = L_{\text{av}} N_{\text{p}} \ln \frac{Y}{y} \tag{V.31}$$

In this equation y is the flexibility of the polymer segments before swelling and may be regarded as an exclusive polymer property. During swelling of the polymer the flexibility reduces from y to Y, depending on the amount of swelling agent. We assume that the average distance between the ends of the polymer chains is s_0 before swelling and s after swelling. Then, in the case of isotropic swelling, the amorphous chains are stretched by a factor

$$\left[\left(L_{av}N_p+\sum_i n_iN_i\right)\Big/(L_{av}N_p)\right]^{1/3} \tag{V.32}$$

and the relationship between s and s_0 is expressed by

$$s=\left[\left(L_{av}N_p+\sum_i n_iN_i\right)\Big/(L_{av}N_p)\right]^{1/3} s_0 \tag{V.33}$$

We now assume that a stretched chain consists of a fully stretched part of minimum flexibility Y_0 and an unstretched part of flexibility y, of which the distance between the ends is s_x (see Fig. 40). Then the average flexibility of the stretched chain is given by

$$Y=\frac{L_{av}-(s-s_x)}{L_{av}}\,y+\frac{(s-s_x)}{L_{av}}\,Y_0=y-\frac{(s-s_x)}{L_{av}}\,\Delta y \tag{V.34}$$

where

$$\Delta y=y-Y_0\approx y-1$$

In this equation y and Δy are properties of the polymer only and hence are constants.

In order to determine Y we need to know the relationship between s_x and s_0. We therefore consider two extreme cases.

Case 1. Kuhn[60] calculated that the average distance between the ends of a polymer molecule, which is free and undisturbed by the presence of other chains, is equal to the square root of the chain length. Hence

$$s_0=\sqrt{L_{av}}$$

If we assume that this relationship holds also for the unstretched part of a stretched chain, then

$$s_x/\sqrt{[L_{av}-(s-s_x)]}=s_0/\sqrt{L_{av}}$$

or

$$s_x=\frac{s_0^2}{2L_{av}}+s_0\sqrt{\left[1-\frac{s}{L_{av}}+\left(\frac{s_0}{2L_{av}}\right)^2\right]} \tag{V.35}$$

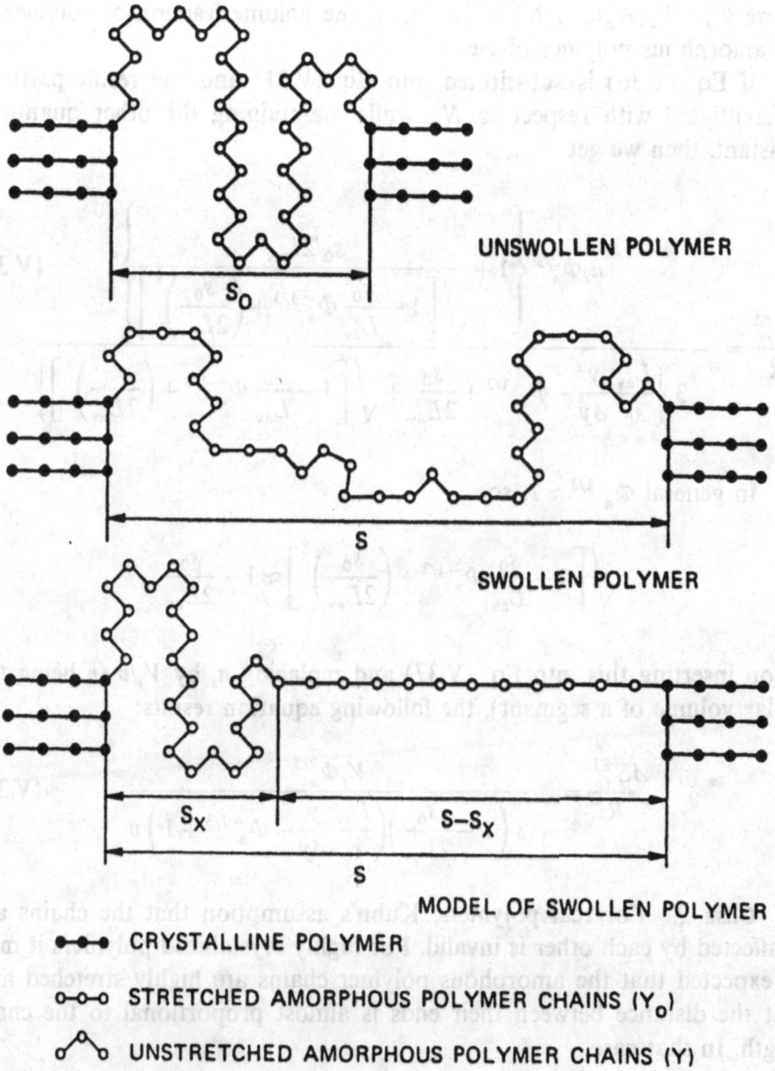

Figure 40. Stretching of amorphous polymer chains by swelling.

Elimination of s and s_x from Eq. (V.34) by means of Eqs. (V.33) and (V.35), respectively, gives

$$\frac{Y}{y} = 1 - \left\{ \frac{s_0}{L_{av}} \Phi_a^{-1/3} - \frac{s_0^2}{2L_{av}^2} - \frac{s_0}{L_{av}} \times \sqrt{\left[1 - \frac{s_0}{L_{av}} \Phi_a^{-1/3} + \left(\frac{s_0}{2L_{av}} \right)^2 \right]} \right\} \frac{\Delta y}{y} \qquad \text{(V.36)}$$

where $\Phi_a = L_{av} N_p/(L_{av} N_p + \sum_i n_i N_i)$ is the volume fraction of polymer in the amorphous polymer phase.

If Eq. (V.36) is substituted into Eq. (V.31) and the result partially differentiated with respect to N_i, while maintaining the other quantities constant, then we get

$$\frac{\Delta S_i^{\mathrm{EL}}}{R} = -\frac{n_i \Phi_a^{2/3}\left\{1 + \dfrac{s_0/2L_{av}}{\sqrt{\left[1 - \dfrac{s_0}{L_{av}}\Phi_a^{-1/3} + \left(\dfrac{s_0}{2L_{av}}\right)^2\right]}}\right\}}{3\left\{\dfrac{L_{av}}{s_0}\dfrac{y}{\Delta y} - \Phi_a^{-1/3} + \dfrac{s_0}{2L_{av}} + \sqrt{\left[1 - \dfrac{s_0}{L_{av}}\Phi_a^{-1/3} + \left(\dfrac{s_0}{2L_{av}}\right)^2\right]}\right\}} \qquad \text{(V.37)}$$

In general $\Phi_a^{-1/3} \approx 1$, so

$$\sqrt{\left[1 - \frac{s_0}{L_{av}}\Phi_a^{-1/3} + \left(\frac{s_0}{2L_{av}}\right)^2\right]} \approx 1 - \frac{s_0}{2L_{av}}$$

Upon inserting this into Eq. (V.37) and replacing n_i by V_i/v (v being the molar volume of a segment), the following equation results:

$$\frac{\Delta S_i^{\mathrm{EL}}}{R} = -\frac{V_i \Phi_a^{2/3}}{3\left(1 - \dfrac{s_0}{2L_{av}}\right)\left(\dfrac{L_{av}}{s_0}\dfrac{y}{\Delta y} - \Phi_a^{-1/3} + 1\right) v} \qquad \text{(V.38)}$$

Case 2. For real polymers, Kuhn's assumption that the chains are unaffected by each other is invalid. For highly crystallized polymers it may be expected that the amorphous polymer chains are highly stretched and that the distance between their ends is almost proportional to the chain length. In that case

$$\frac{s_x}{L_{av} - (s - s_x)} = \frac{s_0}{L_{av}} \qquad \text{(V.39)}$$

Elimination of s and s_x in Eq. (V.34) by means of Eqs. (V.33) and (V.39), respectively, leads to

$$\frac{Y}{y} = 1 - \frac{s_0}{(L_{av} - s_0)}(\Phi_a^{-1/3} - 1)\frac{\Delta y}{y} \qquad \text{(V.40)}$$

By introducing this into Eq. (V.31) and partially differentiating with respect to N_i, while keeping the other quantities constant, we get:

$$\frac{\Delta S_i^{\mathrm{EL}}}{R} = -\frac{V_i \Phi_{\mathrm{a}}^{2/3}}{3\left[\left(1-\frac{s_0}{L_{\mathrm{av}}}\right)\frac{L_{\mathrm{av}}}{s_0}\frac{y}{\Delta y} - \Phi_{\mathrm{a}}^{-1/3} + 1\right] v} \tag{V.41}$$

Comparison of Eqs. (V.38) and (V.41) shows a minor difference in the denominator. For most polymers ΔS_i^{EL} will lie somewhere between these equations. If s_0 is small compared to L_{av} the equations become identical, while for $s_0 \approx L_{\mathrm{av}}$, which is the case for highly crystalline polymers, Eq. (V.41) is valid. We therefore propose application of Eq. (V.41).

If the approximation $\Phi_{\mathrm{a}}^{-1/3} \approx 1$ holds, the latter equation can be further simplified to

$$\frac{\Delta S_i^{\mathrm{EL}}}{R} = -\frac{V_i}{C_{\mathrm{L}}}\Phi_{\mathrm{a}}^{2/3} = -\frac{V_i \Phi_{\mathrm{a}}^{2/3}}{3\left(\frac{L_{\mathrm{av}}-s_0}{s_0}\right)\left(\frac{y}{y-1}\right) v} \tag{V.42}$$

It follows that Eq. (V.42) predicts for the partial molar entropy of elastic strain of the polymer chains a proportionality with $\Phi_{\mathrm{a}}^{2/3}$, unlike the Flory–Rehner equation,[59] which predicts a proportionality with $\Phi_{\mathrm{a}}^{1/3}$, due to their assumption that the distribution function of displacement lengths is Gaussian.

Furthermore, Eq. (V.42) is capable of predicting a tremendous effect of strain on the partial molar entropy if the amorphous polymer chains are almost uncoiled (hence $s_0 \approx L_{\mathrm{av}}$), as is the case with highly crystalline polymers and which cannot be described at all by the Flory–Rehner equation.

VI

PARTIAL MOLAR MIXING ENTHALPY OF MULTICOMPONENT MIXTURES

For multicomponent mixtures consisting of s components, the van der Waals a can be written as

$$a = \sum_{i,j=1}^{s} \theta_i \theta_j a_{ij} \tag{VI.1}$$

and the van der Waals b as

$$b = \sum_{i=1}^{s} \theta_i b_i \tag{VI.2}$$

In accordance with van Laar's treatment[65] (see also Section 6.3) the heat of mixing (Δh^{Mix}) is found to be given by

$$\Delta h^{\mathrm{Mix}} = \sum_{i=1}^{s} \left(\theta_i \frac{a_{ii}}{b_i}\right) - \frac{a}{b} \tag{VI.3}$$

Combination of Eqs. (VI.1)–(VI.3) and rearrangement yield

$$\Delta h^{\mathrm{Mix}} = \sum_{i,j=1}^{s} \frac{\theta_i b_i \theta_j b_j \delta_{ij}^2}{b} \tag{VI.4}$$

where

$$\delta_{ij}^2 = \left(\frac{a_{ii}}{b_i^2} + \frac{a_{jj}}{b_j^2} - 2\frac{a_{ij}}{b_i b_j}\right)$$

The partial molar enthalpy of mixing (ΔH_i^{Mix}) is found by partial differentiation with respect to θ_i while keeping the other quantities constant. This leads to the equation

$$\Delta H_i^{\text{Mix}} = \sum_{\substack{j=1\\ j\neq i}}^{s} b_i \frac{\theta_j b_j}{b}\left(1-\frac{\theta_i b_i}{b}\right)\delta_{ij}^2 - \sum_{\substack{j,k=1\\ j\neq k\neq i}}^{s} b_i \frac{\theta_j b_j \theta_k b_k}{b^2}\delta_{jk}^2 \qquad \text{(VI.5)}$$

In this equation each combination (j, k) is counted only once. If b_i is replaced by V_i and $\theta_j b_j/b$ by Φ_j, etc., we obtain

$$\Delta H_i^{\text{Mix}} = V_i\left[\sum_{\substack{j=2\\ j\neq i}}^{s} \Phi_j(1-\Phi_i)\,\delta_{ij}^2 - \sum_{\substack{j,k=1\\ j\neq k\neq i}}^{s} \Phi_j\Phi_k\delta_{jk}^2\right] \qquad \text{(VI.6)}$$

VII

SOLUBILITY EQUATION FOR GASES IN POLYMERS

The modified Flory–Huggins equation (6.18) of Section 6.4, derived in Appendixes V and VI, is also valid for gases, as pointed out in Chapter 1. It was mentioned in Section 6.6 that, due to the low volume concentration of gases in polymers, this equation can be simplified such that fewer parameters are required for an accurate description of the solubility than are needed for the solubility of liquids. This can be shown as follows.

For isothermal and isobaric systems Eq. (6.18) transforms into Eq. (6.19) (see Section 6.6):

$$\ln a_i = \ln\left(\frac{Z_i}{Z_{av}}\Phi_i\right) + 1 - \frac{V_i}{V_{av}}\frac{Z_i}{Z_{av}}\Phi_s + \frac{V_i}{C_L}\Phi_a^{2/3} + \frac{\Delta H_i^{Mix}}{RT} \qquad (6.19)$$

where the partial molar heat of mixing can be expressed by Eq. (6.16), namely

$$\Delta H_i^{Mix} = V_i\left[\sum_{\substack{j=1\\ j\neq i}}^{n}\Phi_j(1-\Phi_i)\,\delta_{ji}^2 - \sum_{\substack{j=1\\ j\neq k\neq i}}^{n}\Phi_j\Phi_k\delta_{jk}^2\right] \qquad (6.16)$$

with δ_{jk}^2 being the binary mixture parameter (see Section 6.3).

According to Eq. (V.42) in Appendix V.2 the elastic strain factor C_L is given by

$$C_L = 3[f(s) - \Phi_a^{-1/3} + 1]v \qquad (VII.1)$$

where

$$f(s)=\frac{(L_{\mathrm{av}}-s_0)}{s_0}\frac{y}{\Delta y} \tag{VII.2}$$

is a polymer property (constant).

For gases and vapors the activity a_i can be replaced by the ratio of the fugacity f_i and the saturation fugacity $f_{\mathrm{sat},i}$. In the case of gases above their critical temperature (T_c) the saturation fugacity is a hypothetical quantity which, near T_c, can be estimated from the saturation pressure–temperature relationship by applying the Clausius–Clapeyron equation.[22] For temperatures far above T_c it must be determined experimentally, e.g., from sorption experiments.

The sorption of gases in polymers is usually expressed by Henry's law, which in fact is valid only for infinitely low gas pressures according to

$${}^{\infty}H_i=\lim_{f_i\to 0}(C_i/f_i) \tag{VII.3}$$

where C_i is expressed in cm^3(STP) per cm^3 of polymer, f_i in bar, and ${}^{\infty}H_i$ in bar^{-1}. Then the pressure approaches the fugacity.

It can be shown that the relationship between C_i and the volume fraction in the amorphous polymer phase Φ_i is expressed by

$$\Phi_i=\left(\frac{C_i}{c}V_i\right)\Big/\left(\phi_{\mathrm{a}}+\frac{C_sV_s}{c}\right)=\frac{V_i}{\phi_{\mathrm{a}}}\frac{C_i}{c}\Phi_{\mathrm{a}} \tag{VII.4}$$

with

$$C_sV_s=\sum_{j=1}^{s}C_jV_j$$

where s is the number of permeants present in the polymer, c is the factor for converting cm^3(STP) into mol, ϕ_{a} is the volume fraction of amorphous polymer before sorption, and Φ_{a} is the fraction of polymer in the amorphous polymer phase after sorption.

If only one component is sorbed and present at very low concentration, then Eqs. (6.16), (VII.1), and (VII.4) reduce, respectively, to

$$\Delta H_i^{\mathrm{Mix}}=V_i\delta_{ip}^2$$

$$C_{\mathrm{L}}=3vf(s)$$

and

$$\Phi_i = \frac{V_i}{\phi_a} \frac{C_i}{c}$$

Furthermore $Z_{av} \to Z_a$ is the coordination number of the amorphous polymer, $\Phi_s = \Phi_i \to 0$, and $V_{av} = V_i$.

When the various quantities are introduced into Eq. (6.19), replacing the activity by the ratio of the fugacity and saturation fugacity, and finally eliminating the fugacity by means of Henry's law, we obtain

$$-\ln f_{sat,i} = \ln\left(\frac{Z_i}{Z_a}\frac{V_i}{\phi_a}\frac{{}^{\infty}H_i}{c}\right) + 1 + \frac{V_i}{3vf(s)} + \frac{V_i}{RT}\delta_{ip}^2 \qquad \text{(VII.5)}$$

or

$$^{\infty}H_i = \frac{K_H}{f_{sat,i}} \exp\left(-\frac{V_i}{RT}\delta_{ip}^2\right) \qquad \text{(VII.6)}$$

where

$$K_H = \frac{Z_a}{Z_i}\frac{\phi_a}{V_i} c \exp\left[-\left(1 + \frac{V_i}{3vf(s)}\right)\right] = \text{constant}$$

mainly dependent on the polymer.

It follows from Eq. (VII.6) that the Henry coefficient is inversely proportional to the saturation pressure of the gas considered, which is confirmed experimentally as shown in Fig. 13.

By combining Eqs. (6.19), (VII.4), and (VII.5) one derives the equation

$$\ln f_i = \ln\left(\frac{Z_a}{Z_{av}}\frac{C_i \Phi_a}{^{\infty}H_i}\right) - \frac{V_i}{V_{av}}\frac{Z_i}{Z_{av}}\Phi_s + \frac{V_i}{3v}\left\{\frac{\Phi_a^{2/3}}{[f(s) - \Phi_a^{-1/3} + 1]} - \frac{1}{f(s)}\right\} + \left(\frac{\Delta H_i^{\mathrm{Mix}}}{RT} - \frac{V_i}{RT}\delta_{ip}^2\right) \qquad \text{(VII.7)}$$

The volume fraction of gases in polymers is generally very small (namely, less than 0.05), so we may set $Z_{av} = Z_a$. Furthermore

$$\Phi_a^n = (1 - \Phi_s)^n \approx (1 - n\Phi_s)$$

Therefore the third term on the right-hand side of Eq. (VII.7) can be transformed into

$$V_i\left(\frac{1-2f(s)}{9v[f(s)]^2}\right)\Phi_s$$

The last term of Eq. (VII.7) is more complicated. Because very low permeant concentrations are assumed and, furthermore, the interactions between the gases are expected to be small, only the terms in Eq. (6.16) containing Φ_a and $(1-\Phi_i)$ need be taken into account. As a consequence Eq. (6.16) reduces to

$$\Delta H_i^{\text{Mix}} = V_i[(\Phi_s-\Phi_i)(1-\Phi_i)\,\delta_{is}^2+\Phi_a(1-\Phi_i)\,\delta_{ip}^2-(\Phi_s-\Phi_i)\Phi_a\delta_{ps}^2]$$

where

$$\delta_{is}^2=\sum_{\substack{j=1\\ j\neq a\neq i}}^{s}\Phi_j\delta_{ji}^2 \quad\text{and}\quad \delta_{ps}^2=\sum_{\substack{j=1\\ j\neq a\neq i}}^{s}\Phi_j\delta_{jp}^2$$

This equation can be expressed in the form

$$\Delta H_i^{\text{Mix}} = V_i[\Phi_a^2\delta_{ip}^2+\Phi_a(\Phi_s-\Phi_i)(\delta_{ip}^2-\delta_{ps}^2+\delta_{is}^2)] \tag{VII.8}$$

ignoring the term $(\Phi_s-\Phi_i)^2$:

If we assume that $(\delta_{ps}^2-\delta_{is}^2)=\delta_{ip}^2$, then the second term of Eq. (VII.8) vanishes and the last term of Eq. (VII.7) becomes

$$\frac{V_i}{RT}(\Phi_a^2-1)\,\delta_{ip}^2\approx -2\frac{V_i}{RT}\Phi_s\delta_{ip}^2$$

When the various approximations are substituted into Eq. (VII.7), after rearranging the terms we obtain

$$C_i=\frac{f_i^{\infty}H_i}{\Phi_a}\exp(B_i\Phi_a) \tag{VII.9}$$

where

$$B_i=V_i\left\{\frac{[2f(s)-1]}{9v[f(s)]^2}+\frac{1}{V_{av}}\frac{Z_i}{Z_a}+2\frac{\delta_{ip}^2}{RT}\right\}$$

VIII

EVALUATION AND ESTIMATION OF SWELLING PARAMETERS

VIII.1. CRYSTALLINITY

According to Natta the crystallinity of polyethylene is expressed by

$$X_c = \frac{V_a - V}{V_a - V_c} \tag{VIII.1}$$

where $V_a = 1.152 + 6.75 \times 10^{-4} t$ (t in °C) and $V_c = 1.025 + 3.00 \times 10^{-4} t$, X_c being the weight fraction of the crystalline polymer, V_a, V_c, and V the specific volume of the amorphous, crystalline, and total polymer, respectively, while t is the temperature.

The crystallinity of polypropylene is given by

$$X_c = \frac{0.983 + (t + 180) \times 9 \times 10^{-4} - V}{(t + 180) \times 4.8 \times 10^{-4}} \tag{VIII.2}$$

At the swelling temperature of 60°C the density of the polyethylene and polypropylene amounted to 0.914 and 0.863, respectively, corresponding to a crystallinity of 0.660 and 0.352, respectively. It is assumed that the crystallinity is unaffected by the presence of the solvents.

VIII.2. COORDINATION NUMBER

It was noted in Appendix V that the coordination number of a swollen polymer system depends on composition and therefore a volume average of the coordination numbers of all participating molecules was suggested. Hence

$$Z_{av} = \Phi_a Z_a + \sum_i \Phi_i Z_i \tag{VIII.3}$$

with

$$\Phi_a + \sum_i \Phi_i = 1$$

where Z_a is the coordination number of the amorphous polymer (before swelling).

The coordination number depends on the size of the segment chosen. If we choose the CH_2 group as segment for polyolefins, the coordination number of the crystalline phase is $Z_c = 12$. The amorphous phase is less ordered due to twisting and branching of the chains, which explains its lower density. Therefore, it is reasonable to assume that the coordination number of the amorphous- and crystalline-polymer chains are related by their densities such that

$$Z_a = Z_c d_a / d_c \tag{VIII.4}$$

where d_a and d_c are the densities of the amorphous and crystalline polymer, respectively.

Equations (VIII.1) and (VIII.2) were employed to calculate the densities of the amorphous and crystalline phases for polyethylene and polypropylene at 60°C. With Eq. (VIII.4) they yielded $Z_a = 10.5$ for polyethylene and $Z_a = 10.85$ for polypropylene.

It follows from Eq. (6.18) that it is unnecessary to determine the absolute value of the coordination numbers; it suffices to know the ratios of the coordination numbers of the permeants and the polymer (Z_i/Z_a) in order to compute the equilibrium swelling. The coordination numbers of the permeants depend on their own structure and on that of the polymer, as will be shown below.

For *n*-paraffins, which possess the same structure as polyethylene, the identical coordination number may be expected. However, due to their larger molar volume per CH_2 group they occupy in the swollen polymer system more sites than correspond to their carbon number (n). Hence, defining the polymer segment (CH_2 group) as a site it follows that in the pure permeant such a hypothetical site is surrounded by more direct sites than in the polymer. Obviously, the ratios of the coordination numbers of the *n*-paraffins and polyethylene are expressed by the ratios of their specific volumes.

It can be shown that the molar volumes of the *n*-paraffins at 60°C, presented in Table 2, are described accurately by the empirical equation

$$V_i = (n-2) \times v + 2 \times 2.1 \times v = n \times v + 2 \times 1.1 \times v$$

where V_i is the (partial) molar volume of the permeant and v is the molar volume of a polymer segment. Accordingly, the ratio of the coordination numbers of an *n*-paraffin and polyethylene corresponds to

$$Z_i/Z_a = (n \times v + 2 \times 1.1 \times v)/nv = (n + 2 \times 1.1)/n \qquad \text{(VIII.5)}$$

A reasonably good fit of the swelling data for polyethylene in *n*-paraffins is obtained by using Eq. (VIII.5) in order to estimate Z_i/Z_a.

This equation suggests that each CH_2 group in the *n*-paraffin molecules must be counted once, while the larger CH_3 groups must be counted 2.1 times each. An even better fit is achieved if these "nonfitting" groups are counted 2.9 times, so that

$$Z_i/Z_a = (n + 2 \times 1.9)/n \qquad \text{(VIII.6)}$$

This follows clearly from Table 2, in which the measured equilibrium swelling of polyethylene in various *n*-paraffins is compared with the calculated swelling, using Eq. (VIII.6) to estimate Z_i/Z_a.

For nonparaffinic compounds in general a different relationship may be expected. Let us consider cyclanes first. By comparing the structure of cyclohexane and methylcyclohexane with polyethylene it is concluded that, in these compounds, there are at least two "nonfitting" groups and therefore their coordination numbers are also described by Eq. (VIII.6).

If one examines the rigid, flat benzene ring it is expected that none of the six C-atom groups fit the polyethylene lattice and consequently each of these "nonfitting" groups must be counted 2.9 times. Hence

$$Z_i/Z_a = (6 + 6 \times 1.9)/6 = 2.9$$

It follows from Table 2 that, indeed, perfect agreement between measured and calculated swelling of polyethylene in benzene is obtained with this figure.

n-Alkylbenzenes may be expected to exhibit a better fit in polyethylene with increasing length of the n-alkyl chain. Counting each C-atom group in the n-alkyl chain once then yields

$$Z_i/Z_a = (n + 6 \times 1.9)/n \qquad \text{(VIII.7)}$$

Of course, a branched C-atom group in the alkyl chain must considered to be a "nonfitting" group and counted 2.9 times. Therefore, for isoalkylbenzenes

$$Z_i/Z_a = [n + (6 + n_b) \times 1.9]/n \qquad \text{(VIII.8)}$$

where n_b is the number of branches in the alkyl chain.

If the swelling of polyethylene in ethylbenzene is compared with that in the xylenes, it appears that di-alkylbenzenes show a better fit in the polyethylene lattice than mono-alkylbenzenes. In fact a perfect fit is achieved if, in the case of di-alkylbenzenes, only five "nonfitting" groups are counted for the benzene ring instead of six. Consequently the coordination numbers of di(iso)alkylbenzenes are expressed by

$$Z_i/Z_a = [n + (5 + n_b) \times 1.9]/n \qquad \text{(VIII.9)}$$

Taking into account the various considerations, the coordination numbers of the different hydrocarbons in polyethylene were estimated by applying the appropriate equations. The equilibrium sorptions, which are calculated from these data and compared in Table 2 with the respective measured swelling, show in general excellent confirmation.

Let us now consider the equilibrium swelling of polypropylene in various hydrocarbons. Isooctane (2,2,4-trimethylpentane) resembles the structure of polypropylene, except for the second methyl group at the second position of the main backbone, being a "nonfitting" group. Therefore

$$Z_i/Z_a = (n + 1.9)/n \qquad \text{(VIII.10)}$$

Cyclohexane fits the polypropylene lattice exactly and therefore the density ratio of polypropylene and cyclohexane can be used for estimating Z_i/Z_a. The methylcyclanes, however, show at least one "nonfitting" group, so that for these compounds Eq. (VIII.10) is also valid.

Contrary to polyethylene, the benzene ring shows only two "nonfitting" groups in polypropylene. However, each side-chain C atom on the benzene ring alters the fitting of such a molecule in the lattice of polypropylene unfavorably and hence must be regarded as a "nonfitting"

Table 22. Equations for Calculating the Coordination Numbers

Polymer	Type of solvent	$Z_i/Z_a =$
Polyethylene	*n*-Paraffins	$(n+2\times 1.9)/n$
	Isoparaffins	$[n+(2+n_b)\times 1.9]/n$
	(Methyl)cyclohexane	$(n+2\times 1.9)/n$
	Mono-*n*-alkylbenzenes	$(n+6\times 1.9)/n$
	Mono-isoalkylbenzenes	$[n+(6+n_b)\times 1.9]/n$
	Di-*n*-alkylbenzenes	$(n+5\times 1.9)/n$
	Di-isoalkylbenzenes	$[n+(5+n_b)\times 1.9]/n$
Polypropylene	Isooctane	$(n+1.9)/n$
	Cyclohexane	d_a/d_i
	Methylcyclanes	$(n+1.9)/n$
	Alkylbenzenes	$[n+(n-4)\times 1.9]/n$

group. As a result, the coordination numbers of (iso)alkylbenzenes in polypropylene are expressed by

$$Z_i/Z_a = [n+(n-4)\times 1.9]/n \tag{VIII.11}$$

Table 3 shows excellent agreement between the measured equilibrium swelling in the various hydrocarbons with those calculated, using the coordination numbers estimated according to the group contribution model described previously.

Table 22 presents a survey of the various equations for predicting the coordination numbers of hydrocarbons in polyolefins. It must be kept in mind that these equations are valid only at 60°C. The value for "nonfitting" groups (2.9) varies for different temperatures. In order to predict the ratios of the coordination numbers at other temperatures the density ratios were usually applied in this book, because insufficient data were available for working out the group contribution model.

Table 23. Swelling Parameters of Natural Rubber[a]

	Natural rubber	Isooctane	Tetra	Toluene	Cyclohexanone
Density (g/cm^3)	0.92	0.691	1.596	0.868	0.962
Molar volume (cm^3)	—	165.1	96.5	106	102
Z_i/Z_a	—	0.897	0.879	0.884	0.897
Solubility parameter $(cal./cm^3)^{1/2}$	8.00	6.85	8.6	8.9	9.56

[a] Temperature 30 °C.

The coordination numbers in natural rubber were derived from swelling experiments in the respective pure solvents. The results are given in Table 23.

The coordination numbers for the swelling calculations of the ethanol/water/cellulose acetate system were assumed to be identical, owing to the lack of sufficient data.

VIII.3. PARTIAL MOLAR VOLUME

In view of a lack of sufficient data we used the molar volume instead of the partial molar volume, thus ignoring contraction effects.

With the density of the amorphous phase equal to 0.839 for polyethylene, 0.834 for polypropylene, and 0.92 for natural rubber, the molar volumes of the CH_2 groups were calculated to be 16.70, 16.79, and 15.22 cm^3, respectively.

VIII.4. PARTIAL MOLAR HEAT OF MIXING

Hardly any data are available on the partial molar heat of mixing. We therefore estimated values in most cases from the Hildebrand–Scatchard equation (6.13), extended for solvent mixtures.

The solubility parameters of the various solvents and of natural rubber were obtained from the literature.[58] That of polyethylene was derived from those of *n*-paraffins by extrapolation to infinite molecular weight, resulting in $\delta_p = 8.20$ at 60°C. The solubility parameter of polypropylene was set equal to that of cyclohexane ($\delta_p = 7.73$) because of its optimum swelling in that solvent (see Table 3).

The solubility parameter model could not be applied due to the strong polar interactions involved in the ethanol/water/cellulose acetate system. We therefore derived the binary mixing parameters for ethanol/cellulose acetate and water/cellulose acetate from swelling experiments; that of ethanol/water was derived from activity data. The partial molar enthalpy was calculated from these results using Eq. (6.16).

VIII.5. ACTIVITY OF PERMEANTS AND PERMEANT MIXTURES

The swelling experiments were continued until the weight was constant, so it is reasonable to assume that the swollen polymers were in equilibrium with the permeants or permeant mixtures with which they were in contact.

Therefore, in the case of swelling in single solvents the activity of the penetrants are put equal to unity. The activities of the solvents in the swelling experiments with mixtures were calculated from the mixture compositions by applying Eq. (6.18) with $\Phi_s = 1$; hence $\Phi_a = 0$.

VIII.6. ELASTIC STRAIN FACTOR

The elastic strain factor C_L for polyethylene was derived from the swelling experiment in *n*-decane to be 383, using Eq. (6.18) (see Table 2). Accordingly, we found for polypropylene a value of 453 from the swelling experiment with cyclohexane (see Table 3). It can be deduced from the structures of polyethylene and polypropylene that $(y-1)/y$ equals about 2/3 and 5/8, respectively. Assuming Kuhn's relation

$$s_0 = \sqrt{L_{av}}$$

is correct, the value of L_{av} could be estimated from these figures and hence the respective elastic strain factors. An average amorphous chain length of 5.2 nm was found for polyethylene and 7.35 nm for polypropylene, which is not unreasonable in view of the measured crystal sizes of about 15 nm.[100,101]

For natural rubber we calculated the elastic strain factor from the average chain length $L_{av} = 62$ nm, as determined by Paul.[72] By using Kuhn's square-root relation and estimating $(y-1)/y = 3/8$, it is calculated that $C_L = 1744$.

Finally, the value of C_L for cellulose acetate was derived from the swelling experiment in ethanol to be 81.1.

IX

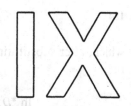

SELF-DIFFUSIVITY IN MULTICOMPONENT MIXTURES

The self-diffusivity of component i in a binary mixture $(i + j)$ is, according to Eq. (8.9) in Section 8.2, given by

$$\ln {}^*D_{i(i+j)} = X_i \ln {}^*D_{ii} + X_j \ln {}^\infty D_{ij} \tag{IX.1}$$

in which X denotes the mole fraction.

Correspondingly, the self-diffusivity of component i in a binary mixture $(i + l)$ is expressed by

$$\ln {}^*D_{i(i+l)} = Y_i \ln {}^*D_{ii} + Y_l \ln {}^\infty D_{il} \tag{IX.2}$$

in which Y denotes the mole fraction as well.

We now consider a mixture of these two binary mixtures consisting of a fraction A_j of the mixture component with j and a fraction A_l of the mixture with component l. By applying the Vigne equation also for this ternary mixture, we obtain

$$\ln {}^*D_{i\mathrm{m}} = A_j \ln {}^*D_{i(i+j)} + A_l \ln {}^*D_{i(i+1)} \tag{IX.3}$$

Combination of these equations yields

$$\ln {}^*D_{i\mathrm{m}} = (A_j X_i + A_l Y_i) \ln {}^*D_{ii} + A_j X_j \ln {}^\infty D_{ij} + A_l Y_l \ln {}^\infty D_{il} \tag{IX.4}$$

which, after substituting $x_i = A_j X_i + A_l Y_i$, $x_j = A_j X_j$, and $x_l = A_l Y_l$, results in

$$\ln {}^*D_{im} = x_i \ln {}^*D_{ii} + x_j \ln {}^\infty D_{ij} + x_l \ln {}^\infty D_{il} \qquad \text{(IX.5)}$$

In the same way a corresponding type of equation can be derived for each number of components s, yielding the general equation

$$\ln {}^*D_{im} = x_i \ln {}^*D_{ii} + \sum_{\substack{j=1 \\ j \neq i}}^{s} x_j \ln {}^\infty D_{ij} \qquad \text{(IX.6)}$$

X

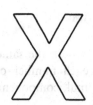

VISCOSITY OF MULTICOMPONENT MIXTURES

The viscosity η_{ij} of a binary mixture $(i + j)$ equals, according to the modified Andrade equation (8.17) in Section 8.4,

$$(k_i X_i + k_j X_j) \ln \eta_{ij} = k_i X_i \ln \eta_i + k_j X_j \ln \eta_j \tag{X.1}$$

where X denotes the mole fraction and k_i is a constant describing the non-ideal mixture behavior of the viscosity.

Correspondingly, the viscosity of a binary mixture $(i + l)$ is given by

$$(k_i Y_i + k_l Y_l) \ln \eta_{il} = k_i Y_i \ln \eta_i + k_l Y_l \ln \eta_l \tag{X.2}$$

where Y is the mole fraction.

We now mix a mole fraction A_j of the mixture with j and a mole fraction A_l of the mixture with l. By applying the modified Andrade equation we obtain

$$[A_j(k_i X_i + k_j X_j) + A_l(k_i Y_i + k_l Y_l)] \ln \eta_m$$
$$= A_j(k_i X_i + k_j X_j) \ln \eta_{ij} + A_l(k_i Y_i + k_l Y_l) \ln \eta_{il}$$

If the binary viscosities on the right-hand side of this equation are substituted by the single-component viscosities with the aid of Eqs. (X.1) and (X.2), then, after rearrangement, we obtain

$$(k_i x_i + k_j x_j + k_l x_l) \ln \eta_m = k_i x_i \ln \eta_i + k_j x_j \ln \eta_j + k_l x_l \ln \eta_l \tag{X.3}$$

where $x_i = A_j X_i + A_l Y_i$, $x_j = A_j X_j$, and $x_l = A_l Y_l$.

In the same way a corresponding type of equation can be derived for each number of components s, yielding the following equation for the multicomponent-mixture viscosity:

$$\ln \eta_{\mathrm{m}} = \sum_{i=1}^{s} (k_i x_i \ln \eta_i) \Big/ \sum_{i=1}^{s} k_i x_i \tag{X.4}$$

XI

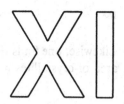

ESTIMATION OF THE GEOMETRIC DIFFUSION RESISTANCE

In the derivation of the equation for the geometric diffusion resistance it is assumed that the permeants dissolve in the amorphous polymer phase only, and that the crystallites remain randomly distributed inside the polymer and do not dissolve by swelling. It is therefore also assumed that the number of "amorphous polymer channels," through which permeation takes place, do not change. On the basis of these assumptions it can be shown that, as a result of swelling, the membrane volume increases by a factor of $(\phi_a + \Phi_a \phi_c)/\Phi_a$, and that the fraction of swollen amorphous polymer phase (polymer + solvent) becomes $\phi_a/(\phi_a + \Phi_a \phi_c)$, where ϕ_a and ϕ_c are the volume fractions of amorphous and crystalline polymer in the original unswollen polymer, respectively, and Φ_a is the volume fraction of polymer in the swollen amorphous polymer phase.

The surface area available for permeation after swelling a unit area of unswollen membrane is defined by

$$A = N \frac{\pi}{4} D_h^2$$

where the hydraulic diameter D_h is given by[102]

$$D_h = 4 \frac{S}{U}$$

By considering infinitely thin slabs it can be proved that the accessible surface fraction S equals the amorphous polymer phase volume fraction, or

$$S = \frac{\phi_a}{\phi_a + \Phi_a \phi_c}$$

Likewise, one finds that the "wetted perimeter" U equals the specific surface area of crystallites in a swollen polymer (M). Now

$$M = \frac{\Phi_a}{\phi_a + \Phi_a \phi_c} M_c$$

where M_c is surface area of crystallites per unit volume of unswollen polymer.

Combination of the above equations leads to

$$A = \frac{4\pi N \phi_a^2}{M_c^2 \Phi_a^2} \tag{XI.1}$$

The relative increase in the total membrane surface area (amorphous + crystalline) during swelling becomes

$$A_s = \left(\frac{\phi_a + \Phi_a \phi_c}{\Phi_a} \right)^n \tag{XI.2}$$

and the relative increase in thickness is given by

$$l/l_0 = \left(\frac{\phi_a + \Phi_a \phi_c}{\Phi_a} \right)^{1-n} \tag{XI.3}$$

where $n = 0$ for fully anisotropic swelling and $n = 2/3$ for isotropic swelling.

Equations (XI.2) and (XI.3) also apply in the case of noncrystalline polymer membranes by putting $\phi_c = 0$. A general expression for the geometric diffusion resistance is

$$R_{\text{geom}} = \int_{z=0}^{l} \frac{dz}{A} = R_0 \int_{z=0}^{l} \Phi_a^m \, dz \tag{XI.4}$$

For noncrystalline polymers $R_0 = 1$ and $m = n$, while in the case of semicrystalline polymers $R_0 = M_c^2/4\pi N \phi_a^2$ and $m = 2$. Quantity R_0 represents the geometric diffusion resistance of a nonswollen membrane per unit area and thickness.

For semicrystalline polymers of high crystallinity a different expression can be derived for R_0. At a random distribution of crystallites the surface area available for permeation equals ϕ_a. If the length of the tortuous "amorphous polymer channels" is t, then

$$R_0 = \frac{t}{\phi_a} \tag{XI.5}$$

If the crystallites are of cubic shape with average length d, the crystallinity amounts to $\phi_c = n_c d^3$, where n_c is the number of crystallites per unit volume of unswollen polymer.

As crystallites and amorphous regions alternate in the polymer, the number of amorphous polymer channels equals the number of crystallites present at the surface per unit area, amounting to approximately $n_c^{2/3}$. If, furthermore, the thickness of the amorphous polymer layers between the crystallites is small relative to d, then the total surface area of the amorphous polymer channels can be approximated by $2dtn_c^{2/3}$, which is also equal to the total surface area of the crystallites $6n_c d^2$. It now follows that

$$t = 3n_c^{1/3} d = 3\phi_c^{1/3}$$

which, in combination with Eq. (XI.5), yields

$$R_0 = 3\frac{\phi_c^{1/3}}{\phi_a} \tag{XI.6}$$

In order to solve the integral of Eq. (XI.4) we define a relative membrane thickness X such that

$$X = \frac{l - z}{l} \tag{XI.7}$$

and set

$$\Phi_a = \Phi_{a,2} - (\Phi_{a,2} - \Phi_{a,1}) X^{\beta_c} \tag{XI.8}$$

If Eqs. (XI.4), (XI.7), and (XI.8) are combined and some straightforward algebra applied, then we obtain

$$R_{geom} = R_0 l \int_{\Phi_{a,1}}^{\Phi_{a,2}} \frac{\Phi_a^m (\Phi_{a,2} - \Phi_a)^{(1-\beta_c)/\beta_c}}{\beta_c (\Phi_{a,2} - \Phi_{a,1})^{1/\beta_c}} \, d\Phi_a \tag{XI.9}$$

which on integration becomes

$$R_{geom} = R_0 l F(\Phi_a)_m \tag{XI.10}$$

where

$$F(\Phi_a)_m = \Phi_{a,2}^m \left(\frac{\Phi_{a,1}}{\Phi_{a,2}}\right)^{m+1} \left[1 + \sum_{j=1}^{\infty} \left(1 - \frac{\Phi_{a,1}}{\Phi_{a,2}}\right)^j \prod_{i=1}^{j} \left(\frac{m + 1/\beta_c + i}{1/\beta_c + i}\right)\right] \tag{XI.11}$$

Calculations show that, in the range of interest (approximately $\phi_c \geqslant 0.5$ and $\Phi_s \leqslant 0.5$), Eq. (XI.11) may be approximated, to an accuracy of a few percent, by

$$F(\Phi_a)_m = \Phi^m_{a,av} \tag{XI.12}$$

where

$$\Phi_{a,av} = \int_{X=0}^{1} \Phi_a \, dX = \frac{\beta_c \Phi_{a,2} + \Phi_{a,1}}{1+\beta_c} \tag{XI.13}$$

When the concentration gradient of the permeants inside the membrane is taken into account, then Eq. (XI.3) for the membrane thickness transforms into

$$l = l_0 \int_{X=0}^{1} \left(\frac{\phi_a + \Phi_a \phi_c}{\Phi_a} \right)^{1-n} dX \tag{XI.14}$$

Combination of this equation with Eqs. (XI.7) and (XI.8) enables numerical calculations to be carried out. These show that l can be approximated with sufficient accuracy by

$$l = l_0 \left(\frac{\phi_a + \Phi_{a,av} \phi_c}{\Phi_{a,av}} \right)^{1-n} \tag{XI.15}$$

For noncrystalline polymers this equation reduces to

$$l = l_0 \Phi^{n-1}_{a,av} \tag{XI.16}$$

Recalling that $m=2$ for semicrystalline polymers and $m=n$ for noncrystalline polymers, it follows that by combining Eqs. (XI.10) and (XI.12) with Eq. (XI.15) for semicrystalline polymers

$$R_{geom} = R_0 l_0 (\phi_a + \Phi_{a,av} \phi_c)^{1-n} \Phi^{1+n}_{a,av} \tag{XI.17}$$

and with Eq. (XI.16) for noncrystalline polymers

$$R_{geom} = R_0 l_0 \Phi^{2n-1}_{a,av} \tag{XI.18}$$

XI.1. ESTIMATION OF THE PARAMETER β_c

In order to calculate the parameter β_c we need an extra equation, for which we have chosen

$$J_i R_0 \int_{z=0}^{l} \frac{\Phi^m_a}{\Phi_s} dz = -\int_{a_{i,1}}^{a_{i,2}} {}^*D_{im} \frac{\Phi_i}{\Phi_s} d \ln a_i \tag{XI.19}$$

This equation can readily be derived from Eq. (9.10) in Chapter 9 by dividing both sides by Φ_s. With the aid of

$$\frac{1}{\Phi_s} = \frac{1}{1-\Phi_a} = 1 + \sum_{r=1}^{\infty} \Phi_a^r$$

Eq. (XI.19) can be expressed in the form

$$J_i R_0 \int_{z=0}^{l} \sum_{r=m}^{\infty} \Phi_a^r \, dz = J_i R_0 l \sum_{r=m}^{\infty} F(\Phi_a)_r$$

$$= -\int_{a_{i,1}}^{a_{i,2}} {}^*D_{im} \frac{\Phi_i}{\Phi_s} \, d\ln a_i \qquad \text{(XI.20)}$$

where $F(\Phi_a)_r$ corresponds to Eq. (XI.11).

In general the approximation represented by Eq. (XI.12) is not allowed, however. The right-hand side of Eq. (XI.20) can be solved in the same way as that described for Eq. (9.10) in Chapter 9.

It is often convenient to eliminate $J_i R_0 l$ in Eq. (XI.20). This can be done by dividing it by Eq. (9.10) to yield

$$F(\beta_c) = \frac{\sum_{r=m}^{\infty} F(\Phi_a)_r}{F(\Phi_a)_m} = \frac{\int_{a_{i,1}}^{a_{i,2}} {}^*D_{im} \frac{\Phi_i}{\Phi_s} \, d\ln a_i}{\int_{a_{i,1}}^{a_{i,2}} {}^*D_{im} \Phi_i \, d\ln a_i} \qquad \text{(XI.21)}$$

The right-hand side of this equation can be calculated from the membrane boundary conditions as described earlier. The result enables β_c to be estimated from the left-hand side of Eq. (XI.21) by an iterative procedure.

XI.2. EXAMPLES OF CALCULATING β_c

We consider semicrystalline polymers with $m = 2$. Pervaporation conditions can be described by

$$\Phi_{a,2} = 1$$

$$\Phi_{a,2} - \Phi_{a,1} = \Phi_{s,1}$$

This, in combination with Eqs. (XI.11) and (XI.21), leads to a relation where β_c depends on $\Phi_{s,1}$ only. Numerical values for β_c were calculated

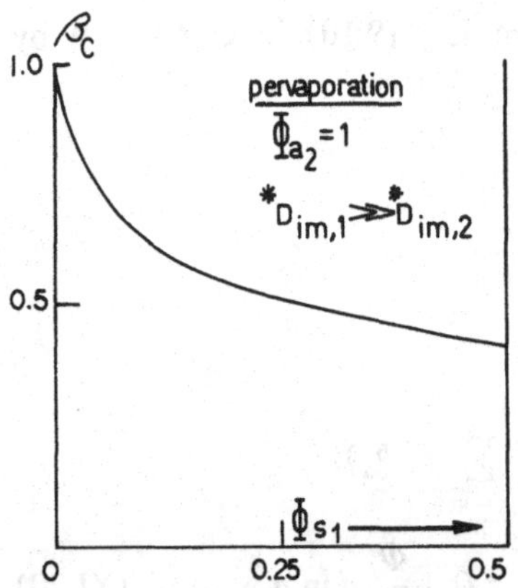

Figure 41. Adjusting parameter β_c depending on swelling.

by the previously mentioned iteration procedure and a graphical representation is shown in Fig. 41.

According to Eq. (XI.11) and the left-hand-side division of Eq. (XI.21), $F(\beta_c)$ is independent of the concentration of the individual permeants and depends only on the total swelling $\Phi_s = 1 - \Phi_a$. Therefore, calculations of $F(\beta_c)$ by means of the right-hand-side division of Eq. (XI.21) should produce identical data whether obtained from component i or j in a binary permeant system. We checked this with results obtained from reverse osmosis experiments with an n-heptane–isooctane mixture (50/50 w/w) through polyethylene at permeation pressures up to 73 bar. Indeed, as shown in Table 24, identical values for $F(\beta_c)$ are found for n-heptane and isooctane.

From these $F(\beta_c)$ values the corresponding β_c values for reverse osmosis conditions were calculated with the aid of Eqs. (XI.11) and (XI.21) and an iteration. The results plotted in Fig. 42 show that β_c decreases with

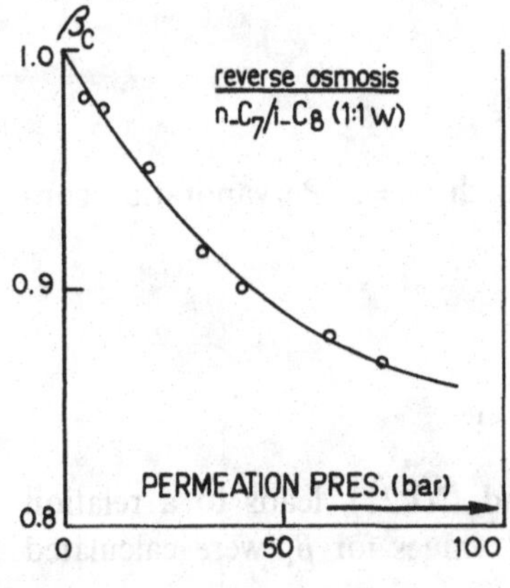

Figure 42. Adjusting parameter β_c depending on permeation pressure.

Table 24. Experimental and Physical Data of the Reverse Osmosis Experiments

Permeation pressure (bar)	Flux (total) ($cm^3/m^2/h$)	Selectivity a_{C_7/C_8}	Equilibrium swelling of the membrane at: upstream side, composition volume fraction, n-C_7	upstream side, composition volume fraction, i-C_8	upstream side, activity, n-C_7	upstream side, activity, i-C_8	downstream side, composition volume fraction, n-C_7	downstream side, composition volume fraction, i-C_8	downstream side, activity, n-C_7	downstream side, activity, i-C_8
5.5	11.1	1.055	0.1419	0.0755	0.5796	0.4427	0.1369	0.0683	0.5731	0.4140
10.0	22.7	1.073	0.1419	0.0755	0.5796	0.4427	0.1315	0.0640	0.5615	0.3978
20.0	49.8	1.144	0.1419	0.0755	0.5796	0.4427	0.1221	0.0548	0.5419	0.3585
32.0	65.4	1.170	0.1419	0.0755	0.5796	0.4427	0.1100	0.0473	0.5086	0.3261
41.0	80.5	1.213	0.1419	0.0755	0.5796	0.4427	0.1029	0.0421	0.4883	0.3004
61.0	93.7	1.300	0.1419	0.0755	0.5796	0.4427	0.0892	0.0331	0.4439	0.2517
73.0	104.0	1.327	0.1419	0.0755	0.5796	0.4427	0.0816	0.0292	0.4160	0.2290

Permeation pressure (bar)	$F(\beta_c)$ calculated for n-C_7	$F(\beta_c)$ calculated for i-C_8	β_c	l/l_0
5.5	4.886	4.886	0.982	1.083
10	5.003	5.002	0.977	1.081
20	5.241	5.239	0.952	1.076
32	5.529	5.524	0.917	1.072
41	5.733	5.725	0.901	1.069
61	6.169	6.154	0.881	1.065
73	6.432	6.412	0.870	1.062

increasing permeation pressure. Obviously, the overall concentration gradient of the permeants inside the membrane deviates more from linearity if the pressure gradient, hence the concentration gradient, increases. At equal upstream permeant content the parameter β_c coincides for reverse osmosis at infinite permeation pressure with that of pervaporation at absolute vacuum on the downstream side.

XII

APPROXIMATE SOLUTION OF THE GENERAL DIFFERENTIAL EQUATION OF MEMBRANE PERMEATION

According to Eq. (9.10) in Chapter 9, the basic diffusivity equation reads

$$J_i R_{\text{geom}} = -\int_{\mu_{i1}}^{\mu_{i2}} {}^*D_{i\text{m}} \Phi_i \left(d \ln a_i + \frac{V_i}{RT} dP \right) \tag{XII.1}$$

With the aid of Eq. (9.8) we obtain

$$ {}^*D_{i\text{m}} = {}^*D_{i\text{m}2} \left[1 + \frac{{}^*D_{i\text{m}1} - {}^*D_{i\text{m}2}}{{}^*D_{i\text{m}2}} f(z) \right]$$

$$\Phi_i = \Phi_{i2} \left[1 + \frac{\Phi_{i1} - \Phi_{i2}}{\Phi_{i2}} f(z) \right]$$

and

$$d \ln a_i = d \ln \left[1 + \frac{a_{i1} - a_{i2}}{a_{i2}} f(z) \right] \tag{XII.2}$$

where[4]

$$f(z) = \left(1 - \frac{z}{l} \right)^{\beta_c} \tag{XII.3}$$

and β_c is the exponent of the exponential concentration gradient in the membrane.

We further assume that, inside the membrane, an exponential pressure gradient also exists such that

$$P = P_2\left[1 + \frac{P_1 - P_2}{P_2} f(y)\right] \tag{XII.4}$$

where

$$f(y) = \left(1 - \frac{z}{l}\right)^{\beta_p} \tag{XII.5}$$

and β_p is the exponent of the exponential pressure gradient inside the membrane.

It can be derived from Eqs. (XII.3) and (XII.5) that

$$f(y) = f(z)^{\beta_p/\beta_c} \tag{XII.6}$$

Elimination of $f(y)$ from Eq. (XII.4) by means of Eq. (XII.6) and elimination of $^*D_{im}$, Φ_i, $\ln a_i$, and P in Eq. (XII.1) by means of Eqs. (XII.2) and (XII.4) result in a differential equation, which can be solved analytically to yield

$$\begin{aligned} J_i R_{\text{geom}} &= {}_{a}D_{im}\, {}_{a}\Phi_i \ln\left(\frac{a_{i1}}{a_{i2}}\right) + {}_{a}\Phi_i\, \Delta^*D_{im} + {}^*D_{im,\text{av}}\, \Delta\Phi_i \\ &+ \frac{V_i}{RT}\Delta P\left[{}^*D_{im1}\Phi_{i1} - \frac{\beta_c}{\beta_p + \beta_c}(\Delta^*D_{im}\Phi_{i2} + {}^*D_{im2}\,\Delta\Phi_i)\right. \\ &\left. - \frac{2\beta_c}{\beta_p + 2\beta_c}(\Delta^*D_{im}\,\Delta\Phi_i)\right] \end{aligned} \tag{XII.7}$$

where

$${}_{a}D_i = \frac{a_{i1}{}^*D_{im2} - a_{i2}{}^*D_{im1}}{a_{i1} - a_{i2}}, \qquad {}_{a}\Phi_i = \frac{a_{i1}\Phi_{i2} - a_{i2}\Phi_{i1}}{a_{i1} - a_{i2}}$$

$$\Delta^*D_{im} = {}^*D_{im1} - {}^*D_{im2}, \qquad {}^*D_{im,\text{av}} = ({}^*D_{im1} + {}^*D_{im2})/2$$

$$\Delta\Phi_i = \Phi_{i1} - \Phi_{i2}, \qquad \Delta P = {}_{m}P_1 - {}_{m}P_2$$

XIII

DERIVATION OF THE GAS PERMEATION EQUATION

In order to derive the gas permeability equation we return to the basic diffusivity equation (9.10) in Chapter 9. For isobaric, isothermal systems this equation is

$$J_i R_{\text{geom}} = -\int_{a_{i1}}^{a_{i2}} {}^*D_{\text{im}} \Phi_i \, d\ln a_i \tag{XIII.1}$$

In Section 9.7 the next relationship is derived for gas sorption:

$$d\ln a_i = d\ln \Phi_i - B_i \, d\Phi_s \tag{XIII.2}$$

which permits elimination of the activity a_i in Eq. (XIII.1).

For calculating the diffusivity in polymers it is convenient to use volume fractions instead of mole fractions.

In general no data are available for the constants k_i in the modified Vigne equation (8.21) for multicomponent mixtures. Furthermore, it may be expected that these values are almost identical for the different gases. Therefore, we apply Eq. (8.10) in Chapter 8 that, for volume concentrations, reads

$$\ln {}^*D_{im} = \Phi_i \ln {}^*D_{ii} + \sum_{\substack{j=1 \\ j \neq i}}^{s} \Phi_j \ln {}^{\infty}D_{ij} \tag{XIII.3}$$

This equation can also be written as

$${}^*D_{im} = {}^{\infty}D_{ip} \exp(\Theta_i \Phi_i) \prod_{\substack{j=1 \\ j \neq a \neq i}}^{s} \exp(\Theta_j \Phi_j) \tag{XIII.4}$$

where a is the polymer, $\Theta_i = \ln({}^*D_{ii}/{}^{\infty}D_{ip})$, and $\Theta_j = \ln({}^{\infty}D_{ij}/{}^{\infty}D_{ip})$.

According to Eq. (9.8) in Chapter 9 the volume fraction of the individual permeants Φ_j can be expressed by

$$\Phi_j = \Phi_{j2} + \Delta\Phi_j f(z) \tag{XIII.5}$$

where

$$\Delta\Phi_j = \Phi_{j1} - \Phi_{j2}$$

and $f(z)$ is the exponential function describing the concentration gradient of the permeants inside the membrane.

When Eq. (XIII.5) is substituted into Eq. (XIII.4), we can derive the relationship

$$*D_{im} = *D_{im2} \exp\left[\left(\sum_{\substack{j=1 \\ j \neq a}}^{s} \Theta_j \Delta\Phi_j\right) f(z)\right] \tag{XIII.6}$$

Combining Eqs. (XIII.1), (XIII.2), (XIII.5), and (XIII.6) yields the next differential equation:

$$J_i R_{\text{geom}} = *D_{im2} \int_{f(z)=1}^{0} \exp\left[\sum_{\substack{j=1 \\ j \neq a}}^{s} \Theta_j \Delta\Phi_j f(z)\right] \times \{\Delta\Phi_i - B_i \Delta\Phi_s [\Phi_{i2} + \Delta\Phi_i f(z)]\}\, df(z)$$

This equation can be integrated to yield

$$J_i R_{\text{geom}} = *D_{im,\ln} \Delta\Phi_i + \frac{B_i}{\theta} *D_{im,\ln} \Delta\Phi_i \Delta\Phi_s - \frac{B_i}{\theta} (*D_{im1}\Phi_{i1} - *D_{im2}\Phi_{i2}) \Delta\Phi_s \tag{XIII.7}$$

where

$$\theta = \ln(*D_{im1}/*D_{im2}) \quad \text{and} \quad *D_{im,\ln} = \frac{(*D_{im1} - *D_{im2})}{\ln(*D_{im1}/*D_{im2})}$$

It follows from Eq. (XIII.4) that for a single permeant system

$$*D_{im} = {}^{\infty}D_{ip} \exp(\Theta_i \Phi_i)$$

Furthermore, it can be derived that

$$^{*}D_{im,\ln} = {}^{\infty}D_{ip}[\exp(\Theta_i \Phi_{i1}) - \exp(\Theta_i \Phi_{i2})]/\Theta_i \, \Delta\Phi_i$$

$$\theta = \Theta_i \, \Delta\Phi_i \qquad \text{and} \qquad \Delta\Phi_s = \Delta\Phi_i$$

Substitution of these expressions in Eq. (XIII.7) yields an equation for the permeation rate of a single permeant:

$$J_i R_{\text{geom}} = {}^{\infty}D_{ip}\left[\frac{\exp(\Theta_i \Phi_{i1}) - \exp(\Theta_i \Phi_{i2})}{\Theta_i}\right]\left(1 + \frac{B_i}{\Theta_i}\right) - \frac{B_i}{\Theta_i}[\Phi_{i1}\exp(\Theta_i \Phi_{i1}) - \Phi_{i2}\exp(\Theta_i \Phi_{i2})] \qquad \text{(XIII.8)}$$

At low concentrations ($\Phi_i \to 0$) $\exp(\Theta_i \Phi_i) \approx (1 + \Theta_i \Phi_i)$, so in this case Eq. (XIII.8) transforms into

$$J_i R_{\text{geom}} = {}^{*}D_{ip}[1 - B_i(\Phi_{i1} + \Phi_{i2})] \, \Delta\Phi_i \qquad \text{(XIII.9)}$$

According to Eq. (9.16) in Chapter 9 the relationship between the fugacity f_i and volume concentration is expressed by

$$f_i = \frac{\phi_a c}{{}^{\infty}H_i V_i} \Phi_i \exp(-B_i \Phi_i) \approx \frac{\phi_a c}{{}^{\infty}H_i V_i} \Phi_i (1 - B_i \Phi_i)$$

Consequently

$$(f_{i1} - f_{i2}) = \frac{\phi_a c}{{}^{\infty}H_i V_i}[\Delta\Phi_i - B_i(\Phi_{i1}^2 - \Phi_{i2}^2)]$$

Combination with Eq. (XIII.9) leads to

$$J_i R_{\text{geom}} = {}^{*}D_{ip} \frac{{}^{\infty}H_i V_i}{\phi_a c}(f_{i1} - f_{i2}) \qquad \text{(XIII.10)}$$

Hence at low concentrations a linear relationship between the permeation rate and the fugacity (pressure) gradient is predicted, while at larger values of B_i a squared dependence on (volume) concentration gradient is predicted.

LIST OF SYMBOLS

A	Cross-sectional area of amorphous phase after swelling of unit area of membrane
A_{ef}	Effective cross-sectional area of a pore
A_0	Real cross-sectional area of a pore
A_s	Total surface area after swelling of unit area of membrane
a	Van der Waals a, Eq. (VI.1)
a	Molecular diameter
${}_{a}D_{im}$	$=\dfrac{a_{i1}\,{}^{*}D_{im2}-a_{i2}\,{}^{*}D_{im1}}{a_{i1}-a_{i2}}$
${}_{a}\Phi_i$	$=\dfrac{a_{i1}\Phi_{i2}-a_{i2}\Phi_{i1}}{a_{i1}-a_{i2}}$
a_i	Activity of component i
a_{ii}	Mutual interaction of molecules i
a_{ij}	Mutual interaction of molecules i with j
a_w	Activity of water
B_i	Parameter, describing nonideal solubility behavior of gases in polymers
b	Van der Waals b, Eq. (VI.2)
b	$=\dfrac{2}{Z_{av}}\,[1-f(0)]\Big/\left[1-\dfrac{2}{Z_{av}}f(0)\right]\approx\dfrac{2}{Z_{av}}$
b_0	$=\dfrac{2}{Z_a}\,[1-f(0)]\Big/\left[1-\dfrac{2}{Z_a}f(0)\right]\approx\dfrac{2}{Z_a}$
b_i	Van der Waals b for component i
b_i	Langmuir affinity coefficient in the dual-mode gas sorption equation
b_1 (b_2)	Backdiffusion factor at upstream (downstream) $= D_0/\delta$
C	Concentration (mol/volume)

C_{bi}	Concentration of component i in the bulk retentate phase
C'_{Hi}	Langmuir capacity constant in the dual-mode gas sorption model
C_i	Concentration of component i (mol/volume)
C_i	Gas concentration [cm^3(STP)/cm^3 of polymer]
C_L	Elastic strain factor, Eq. (V.42)
C_{mi}	Concentration of component i at the interface of the membrane and retentate
C_P	Solute concentration in permeate (mol/volume)
C_{pi}	Concentration of component i in permeate
C_R	Solute concentration in retentate (mol/volume)
C_s	$=\sum C_i=$ total concentration of gases in polymer [cm^3(STP) per cm^3 of polymer]
$C_{s,av}$	Log mean of solute concentration on either side of membrane (mol/volume)
C_w	Water concentration in membrane (mol/volume)
c	Conversion factor for cm^3(STP) into mol (22,400 cm^3/mol)
D	Diffusivity of gases
D_{DD}	Gas diffusivity in bulk polymer phase (dual-mode sorption model)
D_{DH}	Gas diffusivity from bulk polymer phase to microvoids
D_{HD}	Gas diffusivity from microvoids to bulk polymer phase
D_{HH}	Gas diffusivity via microvoids
D_h	Hydraulic diameter
D_i	Fick's diffusivity
${}^{\infty}D_{ia}$	Basic diffusivity of component i in polymer
${}^{*}D_{ii}$	Basic self-diffusivity of component i
D_{ij}	Maxwell–Stefan binary diffusivity
${}^{\infty}D_{ij}$	Basic diffusivity of component i in j
${}^{*}D_{im}$	Self-diffusivity of component i in mixture
D_{im}	Diffusivity of component i in mixture according to Lenoir, Eq. (8.6)
${}^{\infty}D_{im}$	Basic diffusivity of component i in mixture according to Lenoir, Eq. (8.7)
${}^{*}D_{im,av}$	Average self-diffusivity of component i in membrane containing multicomponent permeant mixture
${}^{*}D_{im,ln}$	Logarithmic mean of the self-diffusivity of component i in membrane at the upstream and downstream side
${}^{\infty}D_{mi}$	Basic diffusivity of mixture in component i according to Lenoir, Eq. (8.8)
D_0	DP/P_0, gas diffusivity at standard pressure
D_s	Diffusivity of solute inside membrane

D_w	Diffusivity of water inside membrane
d	Differential
d	Polymer crystal size (edge length of cube)
d_a	Density of amorphous polymer
d_c	Density of crystalline polymer
E^{vap}	Energy for isothermal vaporization of pure liquid to ideal gas state, Eq. (6.13)
F_i	Generalized driving force in Onsager equation
$F(q_i)$	Huggins's correction, Eq. (V.21)
$F(q_p)$	Huggins's correction, Eq. (V.2)
$F(q_s)$	Huggins's correction, Eq. (V.12)
$F(\beta_c)$	See Eq. (XI.21)
$F(\Phi_a)_m$	See Eq. (XI.11)
f	Number of chain ends joining in a crosslink
f_i	Frictional force on component i per mole of i
f_{ij}	Frictional force per mole on i in mixtures with j
$f_{i,av}$	Average frictional force per mole on i in a mixture
$f(0)$	Chance of occupation of a site by a segment of a chain being placed
$f(q)$	See Eq. (V.3)
$f(s)$	See Eq. (6.26)
$f(q_0)$	See Eq. (V.18)
$f(q_{p0})$	See Eq. (V.19)
$f(y)$	Exponential pressure gradient inside membranes, Eq. (XII.5)
$f(z)$	Exponential concentration gradient of permeants inside membranes, Eq. (XII.3)
f_i	Fugacity of component i
$\mathit{f}_{sat,i}$	Saturation fugacity of component i
$^{\infty}H_i$	Henry coefficient (according to the modified Flory–Huggins equation)
i	Component i
IT	Irreversible thermodynamics
J	Total flux through the membrane
J_i	Diffusive flux of component i
J_V	Total volume flux (Kedem–Katchalsky model)
J_s	Molar solute flux
J_w	Molar water flux
K_H	Constant in Eq. (VII.6)
K_i	Permeability of component i
K_w	Permeability of water
$K_{1(2,3)}$	Constants in permeation models, accounting for convective flows in membrane

k	$=k_i/k_j$
k_i	Constant describing nonideal mixture viscosity behavior of component i
k_i	Henry coefficient according to dual-mode sorption model
k_i	D_i/δ = mass transfer coefficient
k_s	Distribution coefficient of solute for liquid mixture and polymer membrane
L	Length of amorphous polymer chain part (expressed as molar volume)
L_{av}	Average amorphous polymer chain length between two crosslinks (expressed as molar volume)
L_{ii}	Straight coefficient in Onsager equation
L_{ik}	Cross coefficient in Onsager equation
L_V	Filtration coefficient (Kedem–Katchalsky model)
l	Thickness of swollen membrane
l_{ij}	Correction for the solubility parameter model of enthalpy
l_0	Thickness of nonswollen membrane
M	Surface area of crystallites per unit volume of swollen polymer
M_c	Surface area of crystallites per unit volume of nonswollen polymer
m	Molar volume ratio of polymer and permeant
m	Exponent in R_{geom}
N	Total mass transfer
N	Number of "amorphous polymer channels" per unit area of unswollen membrane
N_i	Total mass transfer of component i
N_i	Total number of permeant molecules i in swollen polymer phase
N_p	Number of amorphous polymer chains in swollen polymer phase
N_s	Total number of permeant molecules in swollen amorphous polymer phase
n	Parameter describing the degree of anisotropic swelling
n	Number of components, number of different flows, carbon number
n_{av}	Average number of segments per permeant molecule
n_i	Number of segments per molecule i
n_j	Number of collisions of molecule i with molecules j in a liquid mixture
ORR	Onsager reciprocal relation
P	Pressure (for gases at upstream)

$P_{f,av}$	Average gas permeability in membrane
P_i	Partial pressure of component i (at upstream)
P_0	Standard pressure (e.g., 1 bar)
PSCF	Preferential sorption–capillary flow model
p	Downstream pressure
Q^*	Partial molar heat of transfer
q	Variable; also exponent
q_s	qth permeant molecule to be placed in a quasi-lattice
R	Gas constant
R_{geom}	Geometric diffusion resistance of swollen polymer membrane
R_0	Geometric diffusion resistance of nonswollen polymer membrane
R_1 (R_2)	Inner (outer) radius of hollow fiber (tube)
r_p	Equivalent pore radius
S	Cross section per unit membrane area available for mass transport
s	Number of components in multicomponent mixture
s	Average distance between ends of an amorphous polymer chain in a swollen polymer system
s_0	Average distance between ends of a polymer chain in an unswollen polymer system
s_x	See Fig. 40
T	Temperature (K)
T_c	Critical temperature
T_g	Glass transition temperature
t	Temperature (°C)
t	Length of tortuous amorphous polymer channels in semicrystalline polymer membranes
t_w	Critical pore size for retaining solute (see Fig. 2)
U	"Wetted" perimeter per unit membrane area
V	Specific volume of semicrystalline polymer
V_a	Specific volume of amorphous polymer
V_{av}	Average partial molar volume of permeant mixture
V_c	Specific volume of crystalline polymer
V_i	Partial molar volume of component i
V_s	Partial molar volume of solvent
V_s	Average partial molar volume of permeants in polymer
V_t	Total volume of swollen polymer system
V_w	Partial molar volume of water
v	Molar volume of segment (site)
$W(L)\,\delta L$	Distribution function for length of amorphous polymer chains
w_i	Molecular velocity of molecule i

$w_{i,\mathrm{av}}$	Average molecular velocity of molecule i
X	$=z/l$
X_c	Crystallinity (volume fraction)
x_i	Mole fraction of i
x_i	Bulk retentate concentration of i for gases
x_{0i}	Concentration of (gas) i at retentate interface of membrane
Y	Flexibility of polymer segments in swollen polymer system
Y_i	Local permeate concentration of (gas) i
y	Flexibility of polymer segment in unswollen polymer system
y_i	Flexibility of permeant segment
y_i	Bulk permeate concentration of (gas) i
y_0	Flexibility of polymer segment after optimum stretching
y_{0i}	Concentration of (gas) i at permeate interface of membrane
Z	Coordination number
Z'	$=\dfrac{Zm}{(m-1)[1-f(0)]}-\dfrac{2f(0)}{[1-f(0)]}$
Z_a	Coordination number of unswollen polymer phase
Z_{av}	Coordination number of swollen polymer phase
Z_i	Coordination number of pure component i
z	Direction (in polymer) normal to membrane surface
$\alpha^i{}_j$	Selectivity of i relative to j
β_c	Exponent of exponential concentration gradient inside membrane
β_P	Exponent of exponential pressure distribution inside membrane
Γ_i	Activity coefficient of i, based on volume fraction
$\Delta{}^*D_{im}$	$={}^*D_{im1}-{}^*D_{im2}$
$\Delta E_i^{\mathrm{Mix}}$	Partial molar energy of mixing of component i
$\Delta F_i^{\mathrm{Mix}}$	Partial molar free energy of mixing of component i
$\Delta F_s^{\mathrm{Mix}}$	Partial molar free energy of mixing of solvent
$\Delta H_i^{\mathrm{Mix}}$	Partial molar enthalpy of mixing of component i
$\Delta H_s^{\mathrm{Mix}}$	Partial molar enthalpy of mixing of solvent
Δh^{Mix}	Enthalpy of mixing
ΔP	$={}_mP_1-{}_mP_2$
Δs^{EL}	Entropy contribution for elastic strain of polymer chains due to swelling
ΔS_i^{EL}	Partial molar entropy of elastic strain of component i
$\Delta S_i^{\mathrm{Mix}}$	Partial molar entropy of mixing of component i
$\Delta S_s^{\mathrm{Mix}}$	Partial molar entropy of mixing of solvent
Δy	$=y-y_0\approx y-1$
$\Delta\Phi_i$	$=\Phi_{i1}-\Phi_{i2}$

$\Delta\mu_i$	$=\mu_{i1}-\mu_{i2}$
$\Delta\pi_i$	$=\pi_{i1}-\pi_{i2}$
δ	Differential
δ	Thickness of laminar layer (stagnant zone)
$(\delta_a)\ \delta_p$	Solubility parameter of (amorphous) polymer
δ_i	Solubility parameter of component i
δ_{ij}^2	Binary mixture parameter, Eq. (6.17)
δ_s	Solubility parameter of solvent (or mixture)
ε	Fraction of open pore area
ζ_i	Fugacity coefficient of component i
$\zeta_{sat,i}$	Fugacity coefficient of component i at saturation
η_i	Viscosity of component i
η_m	Viscosity of mixture
θ	$\ln(^*D_{im1}/^*D_{im2})$
θ_i	Number of molecules i in a mole mixture
$\Theta\ (\Theta_i)$	$\ln(^*D_{ii}/^{\infty}D_{ip})$
Θ_j	$\ln(^{\infty}D_{ij}/^{\infty}D_{ip})$
Π	Product of terms
π	Osmotic pressure
π_w	Osmotic pressure of water
Φ	Dissipation rate of free energy per unit volume, Eq. (3.1) of Katchalsky and Curran
$\Phi_a\ (\Phi_p)$	Volume fraction of polymer in swollen amorphous polymer phase
$\Phi_{a,av}$	Average volume fraction of polymer in swollen amorphous polymer phase with concentration gradient
Φ_i	Volume fraction of component i in amorphous polymer phase
Φ_s	Volume fraction of solvent in amorphous polymer phase
Φ_s	Total volume fraction of all permeants in amorphous polymer phase (also for gases)
ϕ_a	Volume fraction of amorphous polymer in unswollen semi-crystalline polymer
ϕ_c	Volume fraction of crystalline polymer in unswollen semi-crystalline polymer
χ	Huggins's parameter
χ_0	Temperature-independent fitting parameter of Huggins
μ_i	Thermodynamic potential of component i
μ_w	Thermodynamic potential of water
ρ_i	Density of (gas) i
Ω	Total number of distinguishable configurations of molecules in multicomponent mixture

Ω_{i0} Total number of arrangements in pure phase i
Ω_L Total number of arrangements of all polymer chains of length L in the swollen amorphous polymer phase
Ω_p Total number of arrangements of all polymer chains in the swollen amorphous polymer phase
Ω_{p0} Total number of arrangements of all polymer chains in the unswollen amorphous polymer phase
Ω_q Number of arrangements of the qth polymer chain in the swollen amorphous polymer phase
Ω_{qn} Number of arrangements of the nth segment of the qth polymer chain in the swollen amorphous polymer phase
Ω_{qs} Total number of arrangements of the qth permeant molecule in the swollen amorphous polymer phase
Ω_{qsn} Number of arrangements of the nth segment of the qth permeant molecule in the swollen amorphous polymer phase
Ω_s Total number of arrangements of all permeant molecules in the swollen amorphous polymer phase
ω Solute flux at zero volume flux J_V, Eq. (3.6)
σ Staverman reflection coefficient (coupling coefficient)
σ_h Hydrostatic stress inside membrane
σ_i Size and shape factor of component i
$^*\sigma_{ii}$ Size and shape factor of traced component i in i
σ_{ij} Size and shape factor of component i in a mixture with j
σ_{im} Size and shape factor of component i in a multicomponent mixture
$^\infty\sigma_{ji}$ Size and shape factor of component j infinitely diluted with i
τ Tortuosity

Subscripts

1 and 2 denote upstream and downstream, respectively
l (m) denotes liquid (membrane) phase
0 denotes unswollen state

REFERENCES

1. S. Sourirajan, Reverse osmosis and synthetic membranes; theory, technology, engeneering, National Research Council of Canada, Publication No. NRCC 15627 (1977).
2. M. Soltanieh and W. N. Gill, Reverse osmosis, *Chem. Eng. Commun.* **12**, 279 (1981).
3. Cheng H. Lee, *J. Appl. Polym. Sci.* **19**, 83 (1975).
4. J. G. A. Bitter, *Desalination* **51**(1), 19(1984).
5. R. D. Present and A. J. de Bethune, *Phys. Rev.* **75**(7), 1050 (1949).
6. J. R. Pappenheimer, *Physiol. Rev.* **33**, 387 (1953).
7. N. Lakshminarayanaiah, *Chem. Rev.* **65**(5), 491 (1965).
8. V. P. Budtov, V. P. Vorob'ev, L. A. Gann, and G. D. Myasnikov, *Polym. Sci. USSR* **25**(1), 26 (1983).
9. J. G. A. Bitter, Dehydrogenation using porous ceramic membranes, British Patent Appl. No. 8629135 (December 5, 1986).
10. J. G. A. Bitter and J. P. Haan, Oil/solvent separation with porous membranes, British Patent Appl. No. 8617592 (July 18, 1986).
11. J. G. A. Bitter and M. J. Reynhout, Breaking water–oil emulsions, European Patent No. 0017283 (March 27, 1981).
12. K. Schneider and T. J. van Gassel, Membrandestillation, *Chem.-Ing.-Tech.* **56**(7), 514 (1984).
13. H. L. Fleming, *High Technology*, 22 (August 1987).
14. R. J. R. Uhlhorn, M. H. B. J. Huis in't Veld, K. Keizer, and A. J. Burggraaf, *Membraantechnologie-2* (November 1987).
15. W. van Praag, V. T. Zaspalis, K. Keizer, J. G. van Ommen, J. R. H. Ross, and A. J. Burggraaf, *Membraantechnologie-3* (November/December 1988).
16. J. E. Koresh and A. Sofer, *Sep. Sci. Technol.* **18**(8), 723 (1983).
17. Suzuki Hiroshi, Composite membrane having a surface layer of an ultrathin film of cage-shaped zeolite and processes for production thereof, European Patent Appl. No 0135069 (March 27, 1985).
18. N. N. Li, Liquid surfactant membranes, U.S. Patent No. 3,410,794 (November 12, 1968).
19. A. Kiani, R. R. Bhave, and K. K. Sirkar, *J. Membrane Sci.* **20**, 125 (1984).
20. W. J. Koros, *J. Polym. Sci., Polym. Phys. Ed.* **23**, 1611 (1985).
21. A. C. Puleo, The effects of pendant groups on gas sorption and transport in polymers, Thesis, University of Texas, Austin (December 1988).
22. K. S. Pitzer and L. Brewer, *Thermodynamics* (revision of Lewis and Randall), 2nd ed., McGraw-Hill, New York, 1961.

23. I. Prigogine, *J. Phys. Chem.* **55**, 765 (1951).
24. A. Katchalsky and P. F. Curran, *Non-Equilibrium Thermodynamics in Biophysics*, Harvard University Press, Cambridge, Mass. (1967).
25. L. Onsager, *Phys. Rev.* **37**, 405 (1931); **38**, 2265 (1931).
26. O. Kedem and A. Katchalsky, *Biochem. Biophys. Acta* **27**, 229 (1958).
27. K. S. Spiegler and O. Kedem, *Desalination* **1**, 311 (1966).
28. C. E. Reid and E. J. Breton, *J. Appl. Polym. Sci.* **1**, 133 (1959).
29. S. Sourirajan, *Reverse Osmosis*, Academic Press, New York (1970).
30. H. K. Lonsdale, U. Merten, and R. L. Riley, *J. Appl. Polym. Sci.* **9**, 1341 (1965).
31. T. K. Sherwood, P. L. T. Brian, and R. E. Fisher, *Ind. Eng. Chem., Fundam.* **6**(1), 2 (1967).
32. U. Merten (Ed.), *Desalination by Reverse Osmosis*, MIT Press, Cambridge, Mass. (1966).
33. W. Pusch, Part I, *Ber. Bunsenges., Phys. Chem.* **81**(3), 269 (1977a); Part II, *Ber. Bunsenges., Phys. Chem.* **81**(9), 854 (1977b).
34. G. Johnsson and C. E. Boesen, *Desalination* **17**, 145 (1975).
35. V. T. Stannett, Permeability of plastic films and coatings to gases, vapours and liquids, in *Polymer Science and Technology No. 6* (H. B. Hopfenberg, Ed.), Plenum Press, New York (1974).
36. P. Meares, *J. Am. Chem. Soc.* **76**, 3415 (1954).
37. W. R. Vieth and J. Sladek, *J. Colloid Interface Sci.* **20**, 1014 (1965).
38. W. J. Koros and D. R. Paul, *J. Polym. Sci., Polym. Phys. Ed.* **16**, 1947 (1978).
39. E. S. Sanders and W. J. Koros, *J. Polym. Sci., Polym. Phys. Ed.* **24**, 175 (1986).
40. W. J. Koros, *J. Polym. Sci., Polym. Phys. Ed.* **18**, 981 (1980).
41. J. H. Petropoulos, *J. Polym. Sci.* Pt. A-2, **8**, 1797 (1970).
42. D. R. Paul and W. J. Koros, *J. Polym. Sci., Polym. Phys. Ed.* **14**, 675 (1976).
43. R. M. Barrer, *J. Membrane Sci.* **18**, 25 (1984).
44. E. Sada, H. Kumazawa, P. Xu, and M. Nishigaki, *J. Membrane Sci.* **37**, 165 (1988).
45. J. G. A. Bitter and J. P. Haan, Effect of concentration polarization on membrane separation of gas mixtures, Proc. Int. Cong. on Membranes and Membrane Proc., ICOM'87, Tokyo, Japan, paper 9-OA 1106 (June 8–12, 1987).
46. J. H. Hanemaaijer, Fouling of ultrafiltration membranes; the role of protein adsorption and salt precipitation, Proceedings of "Workshop on Concentration Polarization and Membrane Fouling," University of Twente, The Netherlands (May 18–19, 1987).
47. C. Bootsveld and I. Wienk, Membrane fouling and concentration polarization, Study Soc. of Chem. Tech., "Alembic," Japan Study Tour, University of Twente, The Netherlands (1988).
48. J. M. Dickson, T. Matsumura, P. Blais, and S. Sourirajan, *J. Appl. Polym. Sci.* **19**, 801 (1975); **20**, 1491 (1976).
49. J. G. A. Bitter and H. C. Rijkens, *Membrane Technology 2*, Stam Tijdschriften B.V., The Netherlands (1987).
50. M. L. Huggins, *J. Chem. Phys.* **9**, 440 (1941).
51. P. J. Flory, *J. Chem. Phys.* **9**, 660 (1941).
52. R. W. W. Tock, J. Yu. Cheung, and R. L. Cook, *Sep. Sci.* **9**, 361 (1974).
53. M. H. V. Mulder, T. Franken, and C. A. Smolders, *J. Membrane Sci.* **17**, 289 (1984).
54. R. K. Ghai, H. Ertl, and F. A. L. Dullien, *AIChE J* **19**, 881 (1987).
55. D. W. McCall and D. C. Douglass, *J. Phys. Chem.* **71**(4), 987 (1967).
56. R. C. Reid, J. M. Prausnitz, and T. K. Sherwood, *The Properties of Gases and Liquids*, 3rd ed., McGraw-Hill, New York (1977).
57. J. Crank, *The Mathematics of Diffusion*, 2nd ed., Clarendon Press, Oxford (1975).

58. J. H. Hildebrand and R. L. Scott, *The Solubility of Non-Electrolytes*, 3rd ed., Reinhold, New York (1950).
59. P. J. Flory and J. Rehner, *J. Chem. Phys.* **11**, 512 (1943).
60. W. Kuhn, *Kolloid-Z. Z. Polym.* **68**(1), 1 (1934).
61. P. J. Flory, *J. Chem. Phys.* **18**, 108 (1950).
62. A. Peterlin, *J. Macromol. Sci., Phys.* **B11**(1), 57 (1975).
63. A. S. Michaels and R. B. Parker Jr., *J. Polym. Sci.* **41**, 53 (1959).
64. A. S. Michaels and H. J. Bixler, *J. Polym. Sci.* **50**, 393 (1961).
65. J. J. van Laar, *Z. Phys. Chem.* **72**, 723 (1910).
66. J. D. van der Waals, *Z. Phys. Chem.* **5**, 133 (1890).
67. J. J. van Laar and R. Lorentz, *Z. Anorg. Allg. Chem.* **146**, 42 (1925).
68. C. M. Hansen and A. Beerbower, *Encyclopedia of Chemical Technology, Supplement Volume 1971*, Wiley, New York (1971).
69. W. T. Uragami, H. B. Hopfenberg, W. J. Koros, D. K. Yang, V. T. Stannett, and R. T. Chern, *J. Polym. Sci., Polym. Phys. Ed.* **24**, 779 (1986).
70. J. Brandrup and E. H. Immergut (Eds.), *Polymer Handbook*, 2nd ed., Section V, Wiley, New York (1975).
71. D. G. H. Ballard, A. N. Burgess, J. M. Dekoninck, and E. A. Roberts, *Polymer* **28**(1), 3 (1981).
72. D. R. Paul, J. D. Paciotti, and O. M. Ebra-Lima, *J. Appl. Polym. Sci.* **19**, 1837 (1975).
73. S. A. Stern and A. H. Meringo, *J. Polym. Sci., Polym. Phys. Ed.* **16**, 735 (1978).
74. E. N. da C. Andrade, *Phil. Mag.* (VII) **17**, 497 (1937); *Nature* **125**, 580 (1930).
75. J. Leffler and H. T. Cullinan Jr., *Ind. Eng. Chem., Fundam.* **9**, 84 (1970).
76. J. M. Lenoir, The diffusivity in the liquid state for hydrocarbon mixtures, Fractionation Research Inc., Topical Report No. 87 (July 20, 1982).
77. C. S. Caldwell and A. L. Babb, *J. Phys. Chem.* **60**, 51 (1956).
78. H. T. Cullinan, *Can. J. Chem. Eng.* **46**, 377 (1967).
79. Landolt Bornstein, *Transport Phenomenen (Diffusion in Binaren Flüssigkeiten)*, 6. Auflage, Bd II/5[a], pp. 643, 690 (1969).
80. M. S. Dariel and O. Kedem, *J. Phys. Chem.* **79**(4), 336 (1975).
81. S. Rosenbaum and O. Cotton, *J. Polym. Sci.* Pt. A-1, **7**, 101 (1969).
82. D. R. Paul and O. M. Ebra-Lima, *J. Appl. Polym. Sci.* **14**, 2201 (1970).
83. Y. Shimizu c. s., *J. Polym. Sci.* **17**, 1495 (1979).
84. R. J. Roark and W. C. Young, *Formulas for Stress and Strain*, 504, 5th ed., McGraw-Hill Kogakusha Ltd., Tokyo (1975).
85. M. van der Waal, University of Twente, The Netherlands, private communication.
86. R. F. Madsen, *Hyperfiltration and Ultrafiltration in Plate and Frame Systems*, Elsevier, Amsterdam (1977).
87. M. H. V. Mulder, University of Twente, The Netherlands, private communication.
88. M. H. V. Mulder, J. Oude Hendrikman, H. Hegeman, and A. C. Smolders, *J. Membrane Sci.* **16**, 269 (1983).
89. S. Kimura and T. Nomura, *Maku (Membrane)* **7**(6), 353 (1982).
90. J. S. Chiou and D. R. Paul, *J. Membrane Sci.* **45**, 167 (1989).
91. M. D. Donohue, B. S. Minhas, and S. Y. Lee, *J. Membrane Sci.* **42**, 197 (1989).
92. J. G. A. Bitter and J. P. Haan, European Patent EP 255747 (February 10, 1988); U.S. Patent US 4810366 (March 7, 1989).
93. J. G. A. Bitter, J. P. Haan, and H. C. Rijkens, European Patent Appl. EPO 0220753 (May 6, 1987); U.S. Patent 4748288 (May 31, 1988).
94. J. G. A. Bitter, J. P. Haan, and H. C. Rijkens, Solvent recovery using membranes in

the luboil dewaxing process, AIChE, 1989 Spring Natl. Meeting, Houston, Texas, Proceedings, paper 64D (April 2–6, 1989).

95. J. G. A. Bitter, W. J. M. Weeres, and J. L. W. C. den Boestert, British Patent Appl. No. 8827306 (November 23, 1988).
96. A. van der Scheer and J. Werner, British Patent No. 2144344B (November 11, 1986); U.S. Patent No. 4581043 (April 8, 1986).
97. J. G. A. Bitter and A. J. Schuurmans, British Patent Appl. No. 8827265 (November 22, 1988).
98. R. E. Treybal, *Mass Transfer Operations*, 3rd ed., p. 26, McGraw-Hill, New York (1980).
99. J. O. Hirschfelder, C. F. Curtiss, and R. B. Bird, *Molecular Theory of Gases and Liquids*, Wiley, New York (1964).
100. H. O. Noether and W. Whitney, *Kolloid-Z. Z. Polym.* **251**, 991 (1973).
101. J. Petermann, M. Miles, and H. Gleiter, *J. Macromol. Sci., Phys.* **B12**(3), 393 (1976).
102. P. Grassman, *Physical Principles of Chemical Engineering*, Pergamon Press, New York (1971).

INDEX

Printed in the United States
by Baker & Taylor Publisher Services

Printed in the United States
by Baker & Taylor Publisher Services